ENOUGH!

ENOUGH!

A Modest Political Ecology for an Uncertain Future

MARY LAWHON AND TYLER MCCREARY

agenda
publishing

To Malcolm and Rowan.
And to all of their generation who dream of – and deserve – a better future.

First published in 2023 by Agenda Publishing

Agenda Publishing Limited
PO Box 185
Newcastle upon Tyne
NE20 2DH
www.agendapub.com

ISBN 978-1-78821-620-3

British Library Cataloguing-in-Publication Data
A catalogue record for this book is available from the British Library

Typeset by Newgen Publishing UK
Printed and bound in the UK by 4edge

CONTENTS

ACKNOWLEDGEMENTS

This is, in many ways, a pandemic book, written amid the ongoing uncertainties of school closures, unwell family members, as well as social, political and economic change.

Yet the ideas that underpin it stretch back further and, we hope, have relevance for the world that comes after.

For Mary, the argument in *Enough!* has been shaped by many students, from the University of KwaZulu-Natal to the University of Oklahoma (in particular, Maya Henderson) and most recently the University of Edinburgh. Academics are often told to bring our research into our teaching, but this book is very much the inverse, responding to a collective push to think beyond intellectual genealogies towards a more capacious approach to "what is to be done". Working at UKZN and OU in particular prompted me to develop a framework that could clarify, and at times speak across, environment vs development as well as political and economic divides. Being a parent during the Covid-19 pandemic crystalized a lot of the thinking captured here, and I thank my parents (Joe and LuAnn) and kiddos for enabling me time to write under what were often difficult circumstances. Malcolm, Rowan, you helped me understand both versions of enough, and I promised to thank you in writing for "helping" me type bits of this text. And to Joe, as always.

For Tyler as well, conversations with students have underpinned the development of *Enough!* Working and thinking with students at Florida State University has provided opportunities to consider the possibilities and pitfalls of various proposals to build a just future. It has also challenged me to reach beyond highly polarized political discourse and elicit broader shared hopes and ambitions for a better world. These difficult times have provided an opportunity to re-evaluate my values and priorities, as well as the future that I desire to collectively build with others. I also thank my family and colleagues for their enduring support. Particularly, I owe a debt to Rebecca, who has continually personified the forms of care that enable me to believe that another world is possible and indeed has always been present beneath the surface of capitalist relations.

We would also like to thank James Ferguson, Jamie Peck and Paul Robbins: their work is foundational to our thinking, and supportive conversations both shaped our ideas and gave us the confidence that they were worth pursuing. Cecilia Alda Vidal, Pär Blomkvist, Tim Cresswell, Helena Safron, Erik Swyngedouw, Donald Worster and two anonymous reviewers helped us refine our argument and better define our audience; we appreciate their time, insights and care. Our thinking on modest infrastructure benefitted from supportive conversations with various colleagues at the Beyond Splintering Urbanism workshop in Autun, France, in 2022.

Our thanks as well to Camilla Erskine at Agenda Publishing for her guidance and unfaltering encouragement throughout the writing of this book.

Mary would also like to acknowledge the Riksbankens Jubileumsfond-Swedish Foundation for Humanities and Social Sciences grant "Examining nature-society relations through urban infrastructure" for support for the empirical work in Kampala as well as enabling the book to be published with Open Access. Conversations about modest infrastructure from this project expanded and intersected with ongoing thinking with Tyler about technology, environmentalism and universal basic income in unexpected ways. Thanks to Timos Karpouzoglou, Pär Blomkvist, David Nilsson, Sumit Vij and Gloria Nsangi Nakyagaba. This book, and the reflections on waste in South Africa, were also shaped by the project "Turning livelihoods to rubbish" (with Erik Swyngedouw, Henrik Ernstson and Nate Millington) and associated ongoing writing with Anesu Makina and Kathleen Stokes.

INTRODUCTION

Let us suggest a word for the dominant feeling of the last few years: Enough!

The devastation of Puerto Rico by Hurricane Maria, fears of Day Zero and water shortages in South Africa, the eruption of demonstrations against deaths in police custody and enduring racial injustices in the United States, uprisings against unjust agricultural regulations in India, the global spread of student climate strikes, catastrophic wildfires in Australia, Brazil, the Democratic Republic of Congo and beyond have made clear the political and ecological unsustainability of the present situation.

In 2020, we added the Covid-19 pandemic to the ever-growing list of tragedies, followed by a global cost of living crisis and escalating impacts of climate change.

For many, this continuous sense of catastrophe, and the accompanying fear of worsening conditions, is anything but novel (Davis & Todd 2017; Whyte 2017). The political injunction *¡Ya basta!* rose to prominence a quarter century ago as the central demand of the Zapatistas, an insurgent Indigenous movement in Chiapas, Mexico, rebelling against the erosion of their land rights and self-sufficiency (Khasnabish 2010). The outbreaks of protests and violence in South Africa in 2021 are rooted in decades of frustration over too much injustice and too little change (Africa 2022). Mass disenchantment has also prompted turns to the political right: from Brexit to Bolsonaro, there is a rising current of ultranationalism and increasingly violent xenophobia (Cusset & Wedell 2018; Fraser 2019). For many suffering from the ongoing practices and legacies of colonization, patriarchy, racialization, underdevelopment and other forms of oppression and exclusion, crises have been both enduring and compounding. For many coming of age today in an era of pervasive precarity, life is marked by an overwhelming sense of insecurity and the fear that worse is to come (Lowrey 2022). As Oswin (2019a; see also Oswin 2019b) observes, "There is a spreading exhaustion with the status quo. ... Cries of 'Enough!' can be heard frequently and loudly in all kinds of activist settings".

In this moment, so many of us have clearly had enough.

This frustration, however, is not an endpoint.

Amid all the crises, we can also find hope, desire for change and the seeds of another world.

For these crises are provoking renewed social reflection, re-examination of values and a willingness to consider what was once implausible. Across the globe, the Covid-19 pandemic induced a degree of social solidarity and state intervention in the economy that was truly unimaginable in the years before. As but one example, many governments redistributed money to subdue the worst economic impacts, working alongside an ongoing burgeoning of basic income experiments. The arrival of these new social benefits has spurred demands for their normalization as a common entitlement. From South Africa to Canada and across many places in between, activists and policy-makers are working to solidify and regularize cash transfer programmes beyond the pandemic (Torkelson 2022; Webb 2021; Zhao & Whitehead 2022).

What kinds of change those frustrated with the past and present seek, however, is not always clear, not always agreed upon. It is easy enough to provide critiques of the world we have, to oppose pollution and poverty and name abstract forces like capitalism, racism, nationalism and various forms of kyriarchy[1] that produce inequalities, injustices and unsustainability. It is easy to identify scarcity (hunger, marginalization, disenfranchisement) and surfeit (too much carbon, concentrated wealth, military and police capacity). It is easy to provide partial solutions to particular concerns or describe utopian visions.

We, those of us concerned with injustice and unsustainability, are less clear on our concrete wishes, on the grounds between and beyond more and less, what that means for interconnected, overlapping, and intersecting problems, and how to make pathways forward in our uncertain, imperfect world.

We have thus far failed to provide a radical vision and clear pathway that addresses the inequalities, hierarchies, exclusions and occlusions of our world and is plausible and compelling to a global majority. Myriad visions for change are rooted in radically different understandings of ongoing problems and possibilities. Calls for intersectional analyses and political alliances have, to some extent, blurred old lines. Yet despite numerous and longstanding expressions of solidarity across difference (Sandoval 2000; hooks 2015; Crenshaw 2019), finding common ground among shared struggles continues to be difficult. In practice, social, economic, and ecological concerns often point advocates in opposing and contradictory directions, and draw support from different places and constituencies. What this new politics adds up to remains unclear and difficult to implement.

Into such conversations, we (Mary and Tyler) write *Enough!* (the book). We start with the first meaning of enough: we have had enough! But we also root our thinking in a second meaning: there can be enough for all, and we all deserve enough!

Written in a moment in which many established ways of seeing the world are falling away (Chandler 2018; Fraser 2019; Saad-Filho 2020; Wakefield 2020), this book provides a starting point for considering how our politics might change if it started from the question of "enough". We agree with many scholars and activists who insist that addressing scarcity is a precursor to democratic support for actually existing sustainability, and that addressing surfeit – and particularly concentrations of wealth and power – is necessary for justice and democracy, fairly distributing the material benefits of a sustainable economy.

The approach we take in *Enough!*, however, is rather different from abstract calculations and calls for redistribution, useful as these are. In this sense, we differ substantively from universalizing approaches that seek to determine what enough for all is, then provide it through a centralized authority. Instead, we develop what we call a "modest approach" to the question of enough, one that treats "enough" not as a uniformly knowable metric (a set number of calories, pairs of shoes or kilowatts of energy) but instead as a subjective, variable, political ecological concept.[2] We then consider how our *politics* might change if we considered "enough" as a foundational political ecological claim, and what it would mean to approach this modestly. The politics of enough that we propose here is thus not primarily a call for *more* political action but a different approach to knowledge, nature, science, politics, the state and each other.

This is no small change.

Understanding the shift embedded in a modest politics of enough means re-examining some of the foundations of Western political thought: it means overturning both the centrality of scarcity *and* our confidence about what, and who, can be known and controlled. And then, searching for other options in and beyond the academy.

But wait!

We know: we started this book with a sense of urgency. And now we are asking you, our reader, to *think* more. We agree: this is frustrating, difficult. We too are impatient for change! We too have had enough!

Yet we hope you can bear with us through these pages to see how the thinking we propose here might well enable doing otherwise, facilitating more successful and strategic political efforts.

So let us start this rethinking here: much ongoing modern political and economic theory is based on a strange pairing, a pairing that says the earth is limitless and so are human material desires. In this context, orthodox capitalist and socialist approaches that underpin so much of Western political economic theory argue over the appropriate ownership, regulation and distribution of the means of production in the context of relative scarcity. This scarcity exists, and always will exist, because people always want more. Much environmental theory is based on a partial inversion: that the earth is limited. It retains the assumption that human desires are limitless, and suggests instead that they must be artificially limited. Scarcity, here, exists because people always want more *and* the earth can only provide so much.

What if the assumption that scarcity must always exist is wrong?

What would it mean to invert not only the idea that the earth is limitless, but to *also* reject the idea that material desires are inherently limitless? What would it mean if our approach to justice and sustainability instead started from a recognition that, for most of human history, people did not labour to produce and accumulate surplus: we worked until we had enough, then spent the rest of our days in leisure. This leisure of course was not idle, but probably a mix of restful, creative, artistic, athletic, inquisitive and social activity. This is not a historically controversial position: it is a well-established stance based on new understandings of the long arc of human history (Bregman 2020a; Graeber & Wengrow 2021).

Yet most social theory, written by academics across the north and south, continues to be based on the idea that people always, materially, strive for more, and therefore relative scarcity is a permanent condition.

What if eternally striving for more is actually not an inherent human condition? What might it mean if, instead, we imagined that under some (or even, most) conditions, people are satisfied if they are assured enough, and then seek to spend their time in other ways? What would it mean, for example, to complement the conventional Western understanding of the history of labour as a history of struggle over productivity and gains with a second perspective: as a struggle against systems constructed to entice or coerce people to continue working, even when there was enough for all?

This probably sounds strange and quite possibly very detached from the circumstances of many in the world.

Not so long ago, we (Mary and Tyler) would have thought so too.

If you are sceptical, we are sympathetic.

For now, let us take a step back. Let us note that this is an increasingly common perspective in historical and contemporary studies of work. This is, quite possibly, an increasingly common perspective among ordinary people too as the pandemic has induced a re-evaluation of time, money and things (Gulati 2022). We return to this later in the book. For now, let us pause the macro arc of our argument for a moment and turn to a pair of smaller stories, stories that have shaped how we (Mary and Tyler) came to think about the limits of existing options, and the importance of enough. Let us tell you the stories of Chris and Sipho.[3]

Chris, Sipho and the limits of hegemonic solutions

Chris works in the remote wilderness of the Canadian north, doing environmental monitoring for a transnational pipeline company. He is a young Indigenous man who grew up in the north and took this position as one of the few employment opportunities that would enable him to remain in the region. Being near home is important to Chris, but there are few jobs on his home reserve. Barely half the population there is either working or seeking a job. The rest have ceased looking for employment because of disability, childcare or family responsibilities, or difficulties finding work. Among those still seeking work, the unemployment rate on his home reserve is more than triple the national average. For most, regular employment requires relocating.

Working on pipeline construction allows Chris to spend time out on the land, while regularly seeing his family. Moreover, it pays relatively well, and the money he makes allows him to financially support his immediate family and even some of his extended relations. Work in the remote northern forest can be simultaneously suffocating and isolating. Work camps are often crowded, with little personal space, and require weeks away from families. Nevertheless, it also presents one of the few employment opportunities that allow people to stay in the region.

Pipeline work, however, remains complicated within many Indigenous communities. There is substantial opposition to pipeline projects from some Indigenous and environmental activists. Chris shares these concerns. He has heard his elders talk about the declining salmon stocks. Once, people spoke about a river so thick with fish in the late summer that a person could walk across it on their backs. Now, the fishery is regularly closed to conserve the few remaining salmon. Similarly, the elders talk about the erosion of historical territories. Chris recognizes these concerns, but while his great-grandparents lived off the traplines, almost no one does that anymore. If he does not want to be dependent (on his family members or social assistance), he needs a job. Pipeline work is imperfect, but it is a job where few exist

What are the real possibilities for livelihoods for young residents of remote Indigenous communities? Some have sought to organize against carbon colonialism and block resource extraction regimes, but there are limited opportunities for other livelihoods on the land. Commercial overfishing on the coast damaged salmon stocks, and climate projections suggest spawning streams may disappear entirely, decimating the few remaining fish. Perhaps ironically, the predominantly white environmentalists who decry resource colonialism also oppose the fur trade, undermining one of the few commercial opportunities for making a living off the land. Some Indigenous leaders have leveraged Aboriginal rights claims and political uncertainties to negotiate impact benefit agreements with pipeline companies, garnering jobs guarantees, project revenue-sharing agreements, project co-ownership, community development funds, one-time payments, as well as commitments to address social and environmental impacts. These agreements, however, normalize the regimes of resource extraction that have eroded Indigenous territories and livelihoods, and threaten to exacerbate an expanding climate crisis.

Are there any good choices here? Radical critiques have pointed to the problems of settler colonialism, racism, capitalism and extractivism that run through this story but often left many of those involved still asking: is it preferable to garner some community gains from an expanding extractive catastrophe beyond community control? Or simply oppose it without clear and viable alternative livelihood strategies? Talking with community leaders and activists, Tyler found advocates of both industry partnership and militant resistance were motivated to build a better future for their families and community. Tyler also heard both sides acknowledge that neither collaboration nor resistance was alone sufficient to build the future they desired for their community.

***

Sipho is a waste picker who works on the streets of one of South Africa's formerly white suburbs. He is a migrant, not legally resident in South Africa, and regularly sends money home to Nomvula, the mother of his children, who lives in a rural area (the city is so expensive!). While many think of waste picking as desperate work, Sipho does not see it this way. He is proud of his labour, enjoys the sense of possibility that arises as he walks through the quiet, sunny streets in the early morning. The day is full of possibility. He will search through many bins, but Sipho also has relationships with some of the people who manage these houses, and they sometimes set aside materials for him to collect. Later in the day, he will be joined by comrades and competitors, at times working together and at times competing for the same items.

This is not an easy life: Sipho usually sleeps rough, has had materials stolen, and is regularly harassed by the police as well as those who drive the garbage trucks. Occasionally he has been pressured to limit where he gathers materials by these actors as well as other waste reclaimers. He wishes he spent more time with his family, lived with more security and could imagine safely and securely growing old. This life is one chosen in a context of restraint, in which there are limited opportunities for education and regular, well-paid waged labour.

What possibilities are there for the millions of waste reclaimers globally who undertake work like Sipho every day? Some have tried to organize and improve this work. Strategies range from joining reclaimers together to improve their ability to bargain with those who purchase recyclable materials to providing uniforms for visibility and safety to defining legitimacy and allocating rights to particular areas. Others have tried to eliminate such work altogether, insisting that this is degrading and dirty work that has no place in modern cities. In this line of thinking, waste should be managed through separation at source with waged and unionized workers collecting materials in bulk, with the help of mechanized vehicles. This strategy would surely provide income to fewer people, taking away livelihoods in a place already rife with unemployment. Further, even those with full-time jobs often struggle to meet the needs of the many who depend on their incomes.

What is to be done? Which of these options are better, and for whom? Like Tyler, Mary repeatedly found well-intentioned advocates of both strategies in her research in and beyond South Africa. Both sides seemed aware of the limits of what they proposed, but constrained to the options of improving or eliminating informality.

***

Some academics, including critical and radical ones, insist it remains beyond our scope to address the question of what should be done, whether for waste pickers in South Africa or pipeline workers in northern Canada. Mary and Tyler both spent many years writing about what we saw and facilitating dialogues, and, for a time, we felt this was enough. As outsiders, it was important to listen, read, listen, read and listen some more. We wrote, contextualizing our explanations of these particular struggles and contexts within a matrix of global connections that linked processes across the world.

This reading, listening, writing and listening again, however, failed to provide straightforward and grounded answers when asked – year after year – by research participants the entwined questions: "what should be done?" and "what should I do?" Our answers similarly failed when asked these questions by countless students across the global north and south, a particularly pressing problem for students living in places where fossil fuels sustain the local economy.

It is, we are clear, not our job to answer these questions *for* others. We work and think from a position of privilege as global northern white scholars, a position that remains no matter how much travelling, listening and reflexivity we undertake. Yet we also write with the time and luxury that come with reasonably secure and well-paid jobs, enabling us to listen, read, reflect, listen, compare, evaluate, contextualize, propose, revise and listen again. We write as hybrids, not of the worlds we write about in our research but informed by them to such an extent we are no longer at home quite anywhere. We thus write and think not just about the story of what is, but as people invested in helping us all create, cultivate and share a collective story of what might be.

In this context, for both of us, not addressing the question "what are we for?" felt increasingly problematic (see Ferguson 2010).

But in both our cases, despite reading and thinking across a wide range of approaches, we found no good already-existing answer.

For, in the world we have, incrementalist approaches that improve work experiences and livelihoods may be better than stringent efforts to eliminate bad work that increase poverty. Yet while we are deeply, deeply sympathetic to calls to improve the conditions of millions of waste reclaimers globally, it is surely not too much to say: in a just world, no one's livelihood should have to come from picking through rubbish (see Lawhon 2020 for longer reflections). And while we are deeply, deeply sympathetic to calls to end extractive industries, it is surely not too much to say: in a just world, no one should have to choose between relying on dwindling resources for subsistence and being uprooted, forced to leave home to make a living.

This is, to be clear, not to say that waste reclaiming or subsistence fishing should not happen: some people derive all kinds of value from both of these activities. It is, to reiterate, that no one should *have to* rely on these for a basic livelihood, nor should anyone have to undertake these activities.

In this context, the already-existing proposed solutions we found were either too partial, too much of an ethical compromise, or holistic but dependent on so many other interlinked changes without plausible pathways towards them.

So we wrote a few articles[4] teasing out particular threads in a more abstract way, on the links between automation, justice and unemployment (Pierce *et al.* 2019), on the relationships between basic income, employment and environmentalism (Lawhon & McCreary 2020), the resonance of basic income with anti-essentialist politics (Lawhon & McCreary 2023), and the possibility for basic income to underwrite postcapitalist futures (Lawhon & McCreary 2020, 2023).

Each of these writings, to an extent, helped us in our search and informs the argument we develop here.

Each, too, ultimately also felt too partial, dependent on so many other interrelated changes.

This book is our attempt to tell a fuller story, a story that is of course also partial and in the making, but that outlines the process of change, the future we seek, and a way of thinking that underpins it.

What are we for?

Close your eyes, if you will (after reading the next sentence, that is). What do you envision when we ask about a good green future?

Do you picture earthen houses? Chicken coops, a ceramic solar cooker and a large wooden table for community dinners? Or a dense modern city with flora up the sides of chrome and glass buildings, with a manicured park and people relaxing on benches as a sleek railway runs by? Which of these is the good green future that is plausible? Desirable? More likely to be realized as a shared democratic vision?

Can and should both types of spaces exist in the future you imagine?

In the years before writing this book, Mary regularly conducted a similar exercise at the start of courses on environmentalism, showing students a series of images and asking which is the good green world we seek. Responses differed, often quite viscerally, something Mary expected. Less expected was how visibly surprised some of the students were at the lack of consensus (see Henderson & Lawhon 2021). While all the students in the classes believed in the need for environmental action, they clearly did not agree on a vision of the world they wanted to inhabit, the world they wanted to build.

Such an exercise is, of course, overly simple: images cannot capture the different politics, practices, policies and processes that produce such places. Yet archetypal images of sustainability are useful shorthand for the visions of "more" and "less" that Robbins (2020a) identifies as underpinning ongoing environmental debates. In their approaches to livelihoods, statecraft and economic organization, dominant radical approaches have split over the question of more or less. They advance competing visions of a just and sustainable future, proposing competing pathways for change.

Robbins (2020a) calls the vision of more ecomodern, based on green luxury, in which there is plenty for all. This world may well be nuclear powered, shiny and metallic, without restraint, and guided by a powerful state.

The vision of less is harder to name, but Robbins (2020a) suggests emergent convergence around the term "degrowth" (as we explain in Chapter 2, we prefer the term "arcadian"). It is based on a green world of limits and sacrifices, strong community ties and localized economies, sceptical of modernity and its promises.

Debates over "more" and "less" often result in a stalemate, with advocates on each side entrenched in the defence of their truths (see responses to Robbins, e.g. Gómez-Baggethun 2020; Huber 2021; Luque-Lora 2021, Paulson 2021). But are these really our only options?

We are sympathetic to the many voices who have, often thoughtfully and with a mix of pragmatism and hope, taken and sought to defend positions advocating for "more" and "less", supporting and rejecting modernity.[5] Many also acknowledge that proposals for green austerity and green luxury are imperfect and require considerable struggle, including convincing global populations of their merits.

There is no easy reconciling these narratives.

Yet, many of us (including Mary and Tyler) remain unconvinced by the political and ecological analysis of either side (see Lawhon *et al.* 2021). For many years, we (Mary and Tyler) have rejected the possibility that modern science and a powerful state can sort out our ongoing socioecological mess. We have also doubted that people will, or need to, choose a life of personal sacrifice. We thought there must be something between these options.

No. Not between.

We thought there must be something *beyond* these options.

This book is written for those who, like us, believe there is another story, a story beyond more or less.

In the chapters that follow, we begin to articulate another vision.

The modest approach we develop has "enough for all" as its sociopolitical goal and treats "enough" as a political ecological question.

What does this mean?

Let us start with a few basic studies of ecological economics. In short, here is what they say: we are finally at a point in history in which enough for all can be sustainably produced (see Latham 2000; Millward-Hopkins *et al.* 2020; Creutzig *et al.* 2022). Broadly, we agree with ecological economic calculations that have determined there can be sufficient calories and kilowatts for a decent and dignified life for every person. Such studies typically involve an enormous amount of number crunching and assumptions about what technologies exist and can be considered sustainable. They also involve, necessarily, a lot of universalizing assumptions in order to make such calculations. We cite them here not as definitive proof that these are the "right" volumes of the "right" forms of production of the "right" goods and services, nor as a guidepost towards a universalized future in which everyone consumes the same amounts. They are, instead, an overarching indication of what might be.

Drawing on and beyond such work, we centre our politics on aligning ecologies, technologies, and economies *towards creating* worlds in which there is enough for all.

Thus, despite hegemonic cultural beliefs and political economic systems saying there is never enough, it is crucial to our argument here to say, to repeat, to insist: this is not ecologically, technologically, true. Shortages, as political ecologists have so often shown in particular cases, are effects of the prevailing political economy, not innate or immutable material realities (Matthaei 1984; Mehta 2010). And, mindful of the fraught politics of population, we can be more confident about the ecological calculations in a context in which global population growth has slowed and is predicted to level off (Robbins & Smith 2017). Talking about limits in such contexts – 2 degrees or 350 ppm – obscures a more complex understanding of boundaries as trade-offs amid risk and uncertainty. Instead of fixed limits, in keeping with the science of uncertainty and most political ecological thought, we emphasize constraints and the possibilities for working within, stretching and rethinking them.

Ecological economics thus helps us to defend our argument as ecologically plausible, but does not tell us what this means or how to get there. While there are fixed amounts of certain material things on the earth, there are no fixed amounts of what can and cannot be produced in a world where technology is always changing. People have shown a remarkable ability to create substitutes for scarce resources, to innovate and adapt, and we consider technology to be key to stretching within constraints. We have concerns with modernist approaches to science and technology that are explored in later chapters, but see continued experimentation with and beyond technological artefacts as essential to enabling ways of living more attuned with our world (Berkhout *et al.* 2004; Escobar 2018). That said, we are not cornucopianists: the possibilities of technology are not immaterial, and thus enough for all is most achievable if excesses are curtailed. That is, we collectively should stop producing some things, and certainly stop producing many things the way we have been.

Enough, then, is about less of some things, more of others, a different way of producing most things and a radically different distribution of them.

Crucial to the argument we develop in this book is an imaginary and approach that underpins these points. For while sufficiency can usefully be abstractly calculated, postcolonial and other critical scholarship has made clear that such universalism bears little relevance for the lived desires and experiences of most people and the diversity of socioecological conditions. To put it simply, most of us probably would not agree to putting everything in a pile and dividing it equally, nor would we want to turn over decisions about our everyday material consumption over to a collective. Justly defining and achieving the actual material contents of "enough for all" is, therefore, elusive: we cannot fully and confidently know what it entails nor whether it is possible.

It is precisely amid this muddle that we situate the heart of our argument: we outline a set of political processes and considerations through which enough for all might be achieved *without* collectively defining its material contents. The modest imaginary we articulate is rooted in an acceptance that we cannot make confident and universal claims, but nor are we without any guidance. Quantitative data can usefully inform our thinking without determining pathways forward. Numbers are not inherently modernist: counting long preceded and exists outside modernity! The position we develop here is not a compromise between the more rigorous positions of modern-more and anti-modern-less, but instead draws from contemporary social and ecological sciences that insist that uncertainty is real, and cannot be overcome, but equally ought not paralyse us, nor shift us into hopelessness or extreme relativism.

In sum, this book outlines a modest approach to the political ecological goal of enough for all.

In what follows, we reach across an assortment of critical literatures to draw out a conceptual framework and political pathways towards what we call a modest imaginary[6] and a modest political ecology of the future. This modest imaginary is not an in between position but lies outside the more/less binary, collapsing the dichotomy between modernity and its inverse. This book is therefore about a particular vision of a just and green world of enough, and the pathway we might take to get there.

Yet we do not present our argument as an abstract case. Instead, throughout *Enough!* we narrate our own journey of learning, unlearning and learning anew, reflecting on assumptions and values that we had to rethink alongside our empirics. We do so with the hope that showing you our struggles might help you to better see where you might agree or disagree, where you might pause and reconsider, where you might work with us to deepen our arguments and build new worlds.

We do not want to get too far ahead of ourselves in this introduction, yet let us share with you our conclusion. Maybe it will help you to know where we are headed, even if as yet you feel on rather shaky ground. We believe the clearest path towards justice and sustainability is a politics of enough that can be achieved through a modest state and a diverse, distributionist economy, including the collective provisioning of some services, underwritten by a global, unconditional universal basic income (UBI). A green transition will, no doubt, create different employment opportunities, but a truly sustainable economy will not offer equivalent wages for many workers, and will reduce many other employment options (Lawhon & McCreary 2020). Thus, there is a need to recognize the tensions inherent to change and to create policies to redistribute the impacts of socioeconomic shifts, minimizing resistance to – and spreading the benefits of – sustainability transitions. Providing economic security to all people in such a world *is* a moral good in itself and *also* provides a crucial strategy for subtending socioecological conflicts, creating possibilities for less antagonistic politics, as well as enabling experimentation with and underwriting of non-capitalist economies.

In the chapters that follow, we make clear that this is not an easy pathway, and consider the politics, policies, processes, contingencies, uncertainties and diversity embedded in such an approach. The argument we develop, as we show throughout, is indebted to, complementary and synergistic with many kindred ways of thinking and ongoing already-existing initiatives. We work to draw together congruences, to name a shared, emergent underlying imaginary and understand points of overlap, convergence and difference. We do not believe ours is the only justifiable programme of action, yet we do work to convince you the reader that the modest approach we begin to chart here is a just and politically plausible pathway towards a radically different, sustainable world.

Our roots, audience, and the royal "we"

The ideas we draw upon in the chapters that follow come from a variety of conversations occurring across a breadth of academic fields and spheres of political organizing. We are informed most by our training in political ecology – the study of relationships between environmental issues and political and economic forces – and critiques of development theory and colonialism. We also draw extensively on literature on community economies and feminist scholarship. Our conceptual building blocks – autonomy, diverse economies, a politics of difference, and basic income – are likely familiar terrain to many readers. We work to advance ongoing conversations by placing what are often parallel debates in conversation with each other, developing unspoken synergies that point us towards pathways for change.

Our ideas are also shaped by our experiences growing up, researching and teaching in politically conservative areas, and taking seriously the limits and critiques of orthodox liberal and radical politics. The arguments that we develop here do not easily map onto old political fault lines: we believe deeply in the need for justice and sustainability to be at the centre of politics, and despite much rhetoric castigating the other side, we believe so do most people. Rather than reject and distance ourselves from those who believe in the freedoms promised by capitalism, we seek to engage them in conversation, and show how the story we tell actually shares – accords deeply with – concerns about freedom. We work to develop arguments that centralize multi-scalar self-determination, a widely shared value across the political left, right and centre, mindful of the reality of, and joy that often comes from, deep entanglements with our world and each other.

Much of the value of *Enough!*, we believe, is that it shakes up old ideas and puts them together in new ways, subtending old conflicts and forging new alliances. As we wrote *Enough!*, however, we continuously grappled with how difficult it is to speak across these lines. With each critique and affirmation, we wondered who might be alienated, not necessarily by the idea or argument – although this might also happen – but by our choice of terms and citations. We make note of this concern and our effort here not as a way to resolve this dilemma: our observation does not resolve this conflict nor does our confession let us off the hook! Instead, we share this struggle with you the reader accompanied by an invitation to read a little capaciously, suspend judgements for a while, and dwell in the possibilities with us as we pursue ideas that might enable unconventional alliances in building just and sustainable worlds.

In this context, let us write a brief note on the royal "we". In the pages that follow, we (Mary and Tyler) regularly invoke a collective audience of kindred spirits. Who are we talking to? Who needs to change, who do we see as our allies in our endeavour? Who do we hope to convince of our argument, to join our stories and struggles? In short, we agree that the problem of the so-called Anthropocene is not "people" writ large but rather the dominant political economic order (Moore 2016; Davis & Todd 2017; Wakefield 2020). Most of us have been born into constrained worlds and are doing the best we can under the circumstances. Guilt and responsibility are not the frames we rely on in our call for action. Instead, we all collectively have a vested interest in building better worlds, and we believe that the politics we describe in the chapters that follow would produce a world that is more compelling to the majority of people. We thus write in an effort to convince the broadest possible audience and build unconventional alliances while still retaining the rigour of our argument. We return to more deeply consider who might be attracted to our argument, and the subjectivities it might help us cultivate, in Chapter 6.

We believe the politics we outline have wide resonance, and need a wide range of activist engagement – extending from state initiatives to redistribute wealth to prefigurative experiments with diverse postcapitalist community economies, stretching from labour movements demanding more just working conditions to radical new sociotechnical arrangements – to cultivate joy-filled, caring, convivial lives. It is true that there is a need for coordination across these activities and more engagement from ordinary people. Our central concern in *Enough!*, however, is not so much the tactics for building a movement as the underlying ideas. This is not to undercut the significance of action. Instead, it is to say that movements for justice and environmentalism are not constrained so much by a lack of interest or action. As we detail further in Chapter 2, we think that what is missing is a clearer unifying analysis and shared narrative, and that making time and providing security is essential for enabling wider action. Rather than suggesting new movements, our hope is to shift the focus, language and approach in ways that help existing movements and create wider alliances, ultimately fostering more capacity for change.

We write mindful of the limits of universalisms, yet the politics we describe here are meant to have relevance across the global north and south: we see this not as a contradiction, but as a language and framework through which to build a politics in which many worlds can flourish. Some pieces of our argument resonate more, have greater relevance, require more change in some places than others. In drawing from examples from North America to Africa, we hope to show our argument's broad roots, resonance, and relevance.

We hope that this book is but the start of inquiries into the many shapes and forms of a modest politics in many places.

Outline of the book

The first chapter, "Polarizing political ecologies of the future", reviews dominant radical imaginaries for change. Drawing on ongoing conversations about political ecological futures, we interrogate ecosocialist modernism and arcadian iterations of degrowth (mindful that not all versions of ecosocialism are modernist nor are all versions of degrowth arcadian). We emphasize their contrasting positions on modernity, and elaborate on the limits of embracing and inverting modernity. This provides grounds for our articulation in the next chapter, and throughout the book, of an alternative: a modest imaginary.

Chapter 2 explains our conceptual framework, one that emphasizes the possibilities for building just and sustainable worlds amid the ongoing ever-presence of uncertainty. Modest is positioned as an alternative that collapses problematic binaries produced through modernity, and draws on common ontological and political assumptions across diverse literatures. It accords with many (but not all) non-Western approaches to nature and society and resonates with many contemporary social and ecological theories. Despite this strong resonance with many established ways of thinking, we argue (with much humility) for the need to develop a distinct new term to help guide our thinking as well as political and material practices. We then identify three core pillars of a modest approach: emplaced sustainability, entangled autonomy and anticipatory guiding. These pillars help us to navigate a crowded academic lexicon, positioning our work in relation to emergent thinking about risk, knowledge and uncertainty.

Throughout *Enough!*, we draw on examples from our thinking on infrastructure and livelihoods. Before writing this book, we thought of these as two distinct areas of research, even when they did at times intersect (such as in Mary's work on waste labour and Tyler's research on pipeline politics). Yet the more we began to imagine forward, the more we realized we were developing a way of thinking that worked *across* these topics, an approach that could also be applied to, for example, housing or food. Thus, throughout this book we use examples from our research, but hope to show that a modest approach has broader relevance.

Before mobilizing our framework to explain what a just, green modest future of enough might entail, we provide a brief interlude on cash transfers and their radical potential as we use this as an example of a modest practice elsewhere in the text. Then, with these broad foundations in the literature and conceptualization of a modest imaginary in place, we advance the idea of a modest approach to the economy (Chapter 3), state (Chapter 4) and livelihoods (Chapter 5). Throughout these chapters, we work to both explain broad contours and how modest economies, states and livelihoods might be created. These chapters might be read in any order and we moved them around several times, for there are concerns relevant to Chapter 3 addressed in Chapters 4 and 5, and vice versa. While we present them independently to give some sense of order to the text, the wider approach we articulate is a holistic one, one that works across boundaries of state, economy and livelihoods. Further, while we consider the state, economy and livelihoods to be important, they are not the only topics which can be considered through a modest approach; in our conclusion we touch on questions of culture, identity and subjectivity but note a whole text could be written with this focus. We therefore see this book as a starting point, and hope others might build on, critique and expand our ideas.

The third chapter begins with a brief outline of the limits of modern and anti-modern approaches to the economy. We draw on scholarship that troubles the hegemony of the categories of capitalism and socialism, instead emphasizing diverse economies and alternative markets with pluralist logics. Thinking beyond the binary of socializing production and marketizing social relations, we advance our argument for a modest approach, shifting away from uniform answers (e.g. privatization and marketization under capitalism; state centralization under socialism; or collectivization under communism), urging instead a framework for thinking about justice in a diverse economy. Drawing on broader literature on anti- and postcapitalism, we argue for the erosion of concentrations of wealth and profit-seeking as well as extractive investments (and their associated political influences) through direct regulation and taxation. In conjunction, we argue for growing non-capitalist relations and economies and the enlivening of myriad economic relations, including not only public services and private goods, but also cooperatives, community trusts, local currencies and a panoply of other community economic experiments. We position a basic income as a key, politically possible strategy through which to disperse existing concentrations of wealth and capital and underwrite new economies. This vision of economic diversity and enlivened economies of reciprocity reflects an agenda for prefigurative practice through which we can build a future we desire to inhabit.

We then address the question of what a modest state might look like and do. Modern imaginaries of the state have underpinned many pursuits of justice. Many (but importantly, not all) socialists explicitly or implicitly identify with a modernist lineage, stressing the importance of capturing state power and creating a strong and knowledgeable democratic state. Others have challenged the idea that a just version of such a state is possible. These critiques often come from writers emphasizing identity politics through considerations of race and gender; they challenge the long history of homogenizing and controlling populations through modern statecraft, particularly problematic concerns for those writing from social and geographical margins. Explicitly anti-modern approaches to the state are less prevalent, but some versions of anarchist,[7] postdevelopment and degrowth thought do reject core components of modernity. Yet these scholars and advocates have struggled to explain how radical change might happen without harnessing the power of the state. Addressing this gap, we develop the idea of modest statecraft, articulating what a modest politics might entail, including the types of demands it might make on the state. This approach accepts the inability of the state to deeply know local ecologies and define the contents of the good life, but also the need for the state to act in ways that underwrite the possibilities for justice and sustainability. We focus on two key aspects of a modest state: multi-scalar regulation to provide protection from social and ecological harm and the redistribution of wealth and incomes to assure basic livelihoods. We emphasize that modest politics go beyond certain policies and entail reworking expectations of environment-state-citizen-economy relationships. Doing so is not an easy process but involves a recognition of ongoing struggles that require engaged citizens and, importantly, a socioeconomic context that enables time for doing democracy.

Carrying the story forward, in Chapter 5 we address modest livelihoods. We first explain the modern approach to livelihoods, including the modern work imaginary. This imaginary is premised on the idea that there is a *moral* value to working and that this moral value means that work ought to be the primary source of a livelihood. We demonstrate the long struggle to instil this imaginary in "unruly" subjects, and problematize ongoing modernist assumptions. Moreover, we stress, with increasing automation, a modest world of enough simply does not need the full labour of all the people it can provide for. In this context, we insist that the modern valorization of work must be unlearned before we can justly engage with the question of livelihoods. We then suggest that a modest approach to livelihoods entails the state provision of a basic income and affordable services. We acknowledge ongoing tensions between the state assurance of basic livelihood and autonomy and that no proposal can fully reconcile this tension. Instead, we work to subtend the concerns of some postcolonial, poststructural and feminist theorists who have demonstrated the limits of direct state provision of material goods by urging the provision of income instead of material goods. Further, a UBI uniquely frees time to participate in democracy and emplaced sustainability. In this context, a basic income from the state and socially and politically embedded, regulated services may well be the swiftest and most politically plausible means through which to justly and sustainably achieve enough for all.

Finally, we conclude with reflections on the possibilities and practices for creating a modest politics of enough, including the ways in which our proposal hinges on the enactment of modest subjectivities. We stress that our approach relies upon a different view of human nature than a modernist vision of rational, self-contained individuals and collectivist notions that fully subsume individual interests and preferences; entangled autonomy means recognizing the inexorability of relationality and enabling choice about which relations to forge and cut. We return to Chris and Sipho here, but note more broadly: what these new subjectivities might be is not ours to say.

While in many ways a modest imaginary of a world of enough is a radical vision, for it is starkly different than hegemonic imaginaries and politics, the more we discussed our ideas in and beyond the academy, the less unconventional our argument began to feel. Many of those we spoke with suggest that the arguments we make here in *Enough!* feel intuitively almost-familiar, resonating with insights and experiences, hopes and dreams across the global north and south. The wager underwriting our proposal is that so many people do, fundamentally, want a pathway out of and beyond this crisis-ridden world. We believe the story we tell here is a plausible one, underpinned by a belief that most of us do want to contribute to and live in just and sustainable worlds, so long as this does not preclude economic security.

A modest politics does not provide a single clear pathway, but instead an approach through which to re-envision what is possible, plausible and strategic in our uncertain and imperfect world. Thus, the politics of enough we propose is an ongoing, iterative and agonistic struggle, but is, we hope, sufficiently clear and compelling to help us name and better understand a world we want to build, a world worth fighting for.

While there is a plethora of books now calling for a transition to more sustainable futures, *Enough!* distinguishes itself in two ways. First, it grounds the call for a green future in deep theoretical concerns with the histories of exclusion and oppression that shaped the colonial world and its afterlives, as well as empirical cases across the global north and south. As such, it centres a politics of distribution and a non-universalizing approach to sufficiency. Second, we have sought shared threads across a diverse body of scholarship, ranging from writings on just transitions, degrowth and ecomarxism to diverse economies and postdevelopment studies. Without repeating or being subsumed by any of these categories of thought, we hope to provide a theoretically rooted option distinct from dominant frames that enables us to better understand our current political and ecological conjuncture – and develop a politics that enables us to move beyond it.

In sum, *Enough!* is radical but, we hope, widely resonant, written in a moment in which the failures of existing answers means only radical responses are realistic. It is a story rooted in the places where we dwell, even on pages where no oil or waste politics can be found. A story that is not utopian and does not aspire to answer everything, intended to spark rather than foreclose conversations, to help us recognize and cultivate kindred spirits. It is our hope that the book captures an emergent sense of possibility and direction, naming and helping us find allies and paths forward as we seek the balance gestured at by Paulo Freire in his insightful conversations with Myles Horton: to provide an outline mindful that we collectively will make the road by walking it together (Bell *et al.* 1990).

1

POLARIZING POLITICAL ECOLOGIES OF THE FUTURE

These days, it is hard not to worry about the environment. The climate crisis appears increasingly imminent to the present, evident in stories of disasters like hurricanes, droughts and wildfires that dominate news coverage. Simultaneously, the slow violence of other accumulations of environmental injustice, such as toxic chemicals, soil contamination and erosion and species extinction, continues relatively unabated (Nixon 2011; Cahill & Pain 2019). The costs of environmental degradation, socially and ecologically, are deeply and profoundly uneven (Mohai *et al.* 2009), but it is impossible to fully escape from a sense that there are looming catastrophes ahead.

Lack of change can no longer reasonably be attributed to a lack of awareness of scientific facts or concern for the environment. It no longer makes sense to see "pro" and "anti" environmentalism as the primary lens through which to understand environmental conflict. Even large companies at the heart of unsustainable industries recognize pervasive concerns with the environment: there has been a notable shift from denying environmental impacts towards shifting blame and slowing change (Walenta 2021). As Jon (2021) describes it, from Tulsa, Oklahoma to Darwin, Australia, environmentalisms are happening everywhere – even if not always under this label (see also Hochschild 2016; Lawhon & Makina 2018; Henderson & Lawhon 2021).

Environmental conflicts today are better understood as debates over *what* and *how* we need to change, as well as who "we" is. Framing environmentalism this way builds on scholarship that insists on seeing struggles over resources, access, dispossession and distribution as part of, not separate from, environmentalism (Guha & Martinez-Alier 2013 [1997]; Martinez-Alier 2002; McCarthy 2002; Coombes *et al.* 2012). Yes, there are actors and constituencies that oppose particular initiatives, and disagree over what policies should be implemented and who should bear the financial cost and responsibility for environmental protection initiatives. Scientific controversies remain over the causal mechanisms of the problems that we face, and what kind of science might best help us through.

And yes, climate denialism remains, but even in the United States where vast sums of money have been spent on propaganda (Brulle & Dunlap 2021; Oreskes & Conway 2010), most people agree that human-induced climate change is happening (Marlon *et al.* 2022).

What has always made sustainable development such a compelling frame is that even those who reject particular policies, positions, values and identities still, on the whole, want to live in a sustainable and developed world. The problem of our time is that there is little agreement regarding who and what needs to change, as well as what sociomaterial worlds are even plausible, possible or desirable.

Environmentalism, then, is faced not with a lack of proposed solutions but their near incoherent abundance. We are offered a bewildering array of products and lifestyle options: we are told to ride bicycles, buy organic and avoid BPA. Others emphasize collective action to reform dominant political configurations, for instance, signing petitions, donating to organizations or participating in campaigns for legislative change. Consciousness-raising efforts seek to change the hearts and minds of citizens through environmental education or environmental justice workshops. Building on such efforts, some seek to foster local economies through initiatives such as creating community gardens or local currencies. Others adopt a militant stance to directly confront extractive regimes, attending or organizing protests, establishing blockades or engaging in industrial sabotage.

How are we (collective) to make sense of the many different types of actions recommended by those who care about nature, the environment, health and well-being? Which are pointed in the same direction, towards a similar future; which pull against or undermine each other? How are we to know what to do, how to navigate these choices? These are the questions that so many students have brought to our (Mary's and Tyler's) classrooms. They, and many other thoughtful, considerate and informed people, continue to struggle to make sense of the different and often conflicting range of suggestions.

It was easier when we thought of environmentalism as sharing scientific facts and simply convincing people to care!

Political ecology: critique, diversity and the future

Controversies over knowledge, responsibility and action are at the core of political ecology, a field of study that emerged in the 1970s as a critique of "apolitical" ways of viewing environmental problems (although it has longer roots). Political ecology began to flourish as an academic field in the decades that followed, providing widespread critiques of mainstream environmental science and politics (Bryant & Bailey 1997; Forsyth 2001; Paulson & Gezon 2005). Initially grounded in Marxist theorizations of rural conflicts in the global south, the field's conceptual and empirical breadth has spread while maintaining its focus on critique (Robbins 2015).

Political ecologists have shown that different environmental solutions are rooted in different narratives about the *social* causes of biophysical problems. Some environmentalists blame individual overconsumption. Some attribute responsibility to the state for intervening too much, others blame the same state for intervening too little. For some, the problem lies in the concentration of capital in massive corporations with capacities for action that exceed many governments. Others stress how environmental destruction disproportionately burdens many of the most marginalized people, and diagnose the issue in relation to the structures of racism, ableism, sexism, (neo)colonialism, and so on (see Lawhon 2021 for a review).

These different explanations matter: they tell us what is wrong and, crucially, point towards what needs to be changed if we are to build the just and sustainable world we want.

These different narratives are not just about "the environment" but are underpinned by contrasting notions of fairness, responsibility and justice, as well as what science can know and how ecological knowledge is created. Some are premised on an individualism that advocates for education and rights, as well as instilling moral virtues. Others believe in a collectivist vision, calling for the overthrow of capitalism and capturing of the state to build a common future based on the establishment of collective values. Still others search for justice through reducing hierarchies, increasing autonomy and delinking from the world capitalist system. These different interpretations of what justice and knowledge are, and how they might be achieved, deeply shape how we understand our world and what might be possible.

As a field of study, political ecology has primarily focused on pointing to the limits of mainstream responses to environmentalism. There is a general agreement across political ecological scholarship that capitalism, racism, sexism and other forms of inequality problematically impact the environment and environmental politics. Political ecologists recognize a need for radical change. There is much less clarity on the future that political ecologists would like to see (Robbins 2020a). Yet in the last decade or so, there has been a push – tentative in some ways but increasingly vocal – to move political ecology (and critical scholarship more generally) out of its comfort zone (Braun 2015; White 2019; Desvallées *et al.* 2022; Dimensions of Political Ecology 2022; see Chapter 2). This book builds on calls for political ecologists to consider not just crisis-ridden futures, but the futures we might want to build.

Modernity and the political ecologies of "more" and "less"

How might we understand different ideas of what we collectively might be for, and where our potential areas of agreement lie? In the remainder of this chapter, we consider the question of modernity, a way of understanding the world that remains one of the core fault lines in environmental thought. Arguments for embracing and rejecting modernity in and beyond environmentalism have long roots and myriad versions, and many words have been written explaining their wider histories and geographies. In Box 1, we provide a brief overview of this literature and cognate terms.

Box 1 Modernity and its derivatives

Modernity is an unwieldy word. Many volumes have been written to explain, critique and defend the term (e.g. Beck *et al.* 1994; Amin 2009; Chakrabarty 2011; Mignolo 2011), and in these short pages we highlight some key contours in order to be able to move beyond them.

We use the term "modernity" to signify an imaginary of what the world is and how it works. We use it in a more narrow sense than many authors; as we return to at the end of this box, this enables us to more sharply identify what we are for and against. Modernity is not a real, empirically knowable and measurable thing. As an idea, however, it has had tremendous impacts on the world, shaping the way people act, what they expect and how they relate to others and the environment. Modernity is difficult to map onto most political spectrums, and advocates for and against modernity write in and through the language of justice, socialism and equality.

Modernity's roots and origins are, of course, impossible to singularly pinpoint, but modernity is generally understood to have been a profoundly radical, revolutionary idea that developed and was consolidated in Europe in the seventeenth century. The modern imaginary is premised on the idea that the world is knowable. This knowledge comes not from (the) God(s) nor from abstract reflection, but from rational experimentation in a mechanistic world. Knowledge is generated through identifying categories of knowable things. In other words, modern science is imagined as the study of discrete objects in pure conditions to understand their properties, then put these objects back together again and know how the world works. Knowing how it works means having the information through which to control (Merchant 1980). Unruly things like nature can, through this science, be made to work more efficiently (for the benefit of certain men/mankind).

These principles are then applied to society. Early modern thinking divided humanity into categories: there were distinct genders and races, and understanding the true properties of each was a core purpose of early social science (Hannah 2000; Curtis 2002). These categories were not neutral: they were hierarchical, with white men at the top of the hierarchy. Of course, even this category was further refined, with some white men placed above others. Those at the top of the hierarchy were deemed rational, intelligent individuals, and therefore it was presumed to be quite reasonable for the world to operate as they deemed fit (Mignolo 2011). Defining the qualities of different people meant that diverse populations, like nature, could be controlled (Mamdani 2012). For some, this control was pursued through violence. Others viewed the hierarchy as temporal, with some people(s) simply behind. These "others" were to be "improved", to be taught and guided to "civility" (this was the original version of the white man's burden). This process was called modernization, and underpinned the emergence of development regimes (Lushaba 2009).

Modernity is a totalizing theoretical frame, one in which all the parts relate to and require others. This was and is part of its appeal. However, modernity's meaning and application also was and continues to be contested. Some use it to defend hierarchies while others use the language of liberal modernity to oppose sexism, racism, colonialism and other injustices.

Many critics of modernity nevertheless identify its totalizing frame as a core fault; one cannot pick and choose certain parts of modernity without, intentionally or not, drawing in others (Cooper 2005). Eschewing modernity is not the same as rejecting science, numbers and quantification. Rather, rejecting modernity more narrowly rebuts the totalizing aspiration to engineer nature and society through the application of modernist scientific methods and designs. Rejecting modernity also does not mean refusing all ideas that emerge in Euro-America. Non-modern ways of understanding the world have always been part of Western thought and politics, and they continue to be so today.

While modernity continues to have its defenders, modernity as we have described it is no longer an explicit, hegemonic narrative even in the global north, nor is it an accurate representation of the world as it is. Modernity was never fully believed by a global majority, and a modern world was never achieved anywhere: despite many efforts, it is simply not possible to make the world we have match the modern idea of it. As Latour (1994) has rather famously and succinctly claimed, "we have never been modern". The assumption that, through science, we can know and control the world and each other has failed to hold up in practice time and again. Projects collapse, people revolt and patchwork fixes are required and hidden to maintain the illusion of modern power, for the world is more complex and uncertain than modernists anticipated. Even in the most modern of moments, in the most modern places, and for the most modern people, modernity remains elusive and distant.

Modernity thus is better understood as an imaginary that informs beliefs and actions rather than an actual condition, an aspiration held by some people, one that has constantly been accompanied by anxieties about its impacts (Everdell 1997). Further, most, in the world we have today, would not publicly defend modern ideas of colonial racial hierarchies, and there are ongoing efforts to expose and rework the modernist assumptions that underpin much ongoing natural and social science. Recognizing modernity as a controversial sociopolitical heuristic, an imaginary rather than descriptor of the world that is, does not reduce its impact: modernity remains a powerful idea that shapes how people interact with each other and our surroundings despite, and maybe because of, its elusivity.

The quest to know and control the world through rational empiricism is, however, not quite over. Some believe that humanity continues to be on a quest which will ultimately provide the answers, enabling knowledge and control. This form of scientific practice, sometimes called postpositivist, accepts that knowledge is partial and seeks to control bias to build the closest approximation to reality (Popper 1978). Rather than presuming that the world can simply be empirically observed, scholars accumulate and ana-lyse increasing volumes of data to calculate and control for uncertainty and improve predictions of social and ecological phenomenon (Hacking 1990). This hubris has been particularly emboldened with aspirations for total knowledge in the era of big data and increased penetration of computer algorithms into everyday life (Kitchin & Dodge 2011; Kitchin 2014). New digital technologies are monumentally expanding the volume of available data and computing power, resulting in gigantic advances in the algorithmic modelling of complex dynamics, often under private control (Srnicek 2017; van Dijck et al. 2018; Zuboff 2019). Some believe that improved knowledge, and growing aspirations for control of data to model populations and their activities, can provide more accurate predictions and underpin the design of social and ecological interventions. A crass, mathematical interpretation of intersectionality is evident in, for instance, analyses of the interplay of race, class, gender, education and religious affiliation in predicting voting patterns (Silver 2012; for a critique, see Benjamin 2019). Even concepts such as resili-ence, originally formulated as a challenge to the engineering mentality that sought to maintain the stability of an existing order (Holling 1973; see

Chapter 2), have been co-opted as a new mode of power operating through processes of "controlled decontrolling" that seek to efficiently manage risk while strategically abandoning forms of life deemed uneconomic or expendable (Lorimer 2020, for example, refers to this as Modernity 2.0).

Derivatives of modernity

Modernity, then, is an imaginary that includes an explanation of how people can come to know our world and each other. Here, we lightly touch on a series of cognate terms and associated literatures, noting that many alternatives to modernity exist but that none quite capture the essence of the argument developed throughout *Enough!* (see also Chapter 2).

In its earlier formulations, modernity was contrasted primarily with "tradition". Tradition could be found globally, including in Europe, and is characterized by slow change, strong kinship, spiritual beliefs and livelihoods that were typically drawn from their local surroundings. While "tradition" existed long before modernity, the invention of tradition as a domain of purportedly unchanging practices emerged as the co-constituted "other" of modernity with its constant change and innovation (Hobsbawm & Ranger 1983). Early social scientists, such as Émile Durkheim (1997 [1893]), expected that modern people would reject tradition as a source of knowledge and replace it with reasoning rooted in empirical science, weak social ties, quickening of change, secular beliefs and urban-industrial livelihoods. However, as industrial change quickened in the modern era, there was a concurrent rise of fundamentalism, promoting nostalgia for a purportedly simpler past and a return to its absolute values (Harvey 1990; Armstrong 2000). While "tradition" continues to hold sway as a political concept, within postcolonial thinking it is widely seen as both a pejorative and an inaccurate representation of non-modern cultures. We do not use this word elsewhere in this book.

Postmodernism as a term emerged in the 1970s, largely in parallel to the critiques of modernity and tradition that permeated colonial thinking, development studies and postcolonial scholarship. Temporally, it emerged "after modernity", and the extent to which postmodern thinking rejects or moves beyond modern thinking is subject to some debate. Some describe it as characterized by scepticism of the universal claims of modernity, moral ambiguity, whimsy, spectacle and a combination of nostalgia for the past and dwelling in the present. Given its ambiguities and rather fraught politics, we limit our use of this word elsewhere in the book.

We consider tradition and postmodern as "anti-modern" terms generated through and in relation to modernity. Both are, like modest, not modern. The modest approach that we develop throughout this book, however, is not defined through an inverse and antagonistic relationship to modernity, but is rather positioned as an alternative to it (see Chapter 2).

Some have argued for reformulating modernity, making the term more capacious, recognizing "multiple modernities" (Eisenstadt 2002; in environmental thought, see White 2019). We are sympathetic to the many who want to be included in the thing called modernity; we also agree with the core premise of multiple modernities scholarship that rejects a single teleological narrative of development. We also agree that there are multiple origins of modern thought beyond Europe: the emergence of Renaissance and Enlightenment modern science, for example, drew extensively on non-European knowledge; modern science emerged and was advanced in the colonies and through colonial relations (Livingstone & Withers 1999; Amin 2009; Mignolo 2011). Modernity is thus not purely a construct of the West, but was produced in relation to other places and peoples.

Yet much of the work on multiple modernities – and other work that emphasizes its plurality – blurs the meaning of the term "modernity", resulting in a term that is so capacious that it is unclear what is *not* modern. "Modernity" then becomes hard to differentiate from other imaginaries, instead becoming a temporal flag for "the present" (Ferguson 2005).

We are fully in agreement that the history of modernity is spatially diffuse and that history is comprised of multiple ways of understanding. And further, not all of the ideas that emerged in Europe over the last few hundred years are modern.

Much of the history of social justice might be understood as contestations over what modernity is as well as between modern and non-modern ways of pursuing justice. There is no simple alignment between modernity and (in)justice. While there are connections between hierarchies, racism, patriarchy, colonialism, environmental degradation and modernity, all of these problems also exist outside of modernity. Moreover, invocations of the need to defend traditional values from modernity often augur revanchist and exclusionary movements. We need to understand how the idea of modernity permeates our history and present through movements both advocating and opposing it.

In this context, we insist on the need – the political and analytical utility – of a narrow meaning of modernity so as to recognize "ideas, movements or societies [that] fall outside its remit" (Fourie 2012). Distinguishing between modern and non-modern ideas within and beyond Western thought is crucial to enabling us to navigate which parts of what histories from which places we ought to embrace and reject. Simply refusing modernity does not necessarily create a just world. Rather we must carefully reckon with modernity and its legacies as we seek to imagine different futures.

In this chapter, we consider how conflicting views on modernity underwrite two prominent versions of environmentalism, generating orthogonal approaches to what is possible, plausible and desirable. Perhaps the most iconic version of radical environmentalism is an arcadian, romantic vision in which people are deeply in touch and live in harmony with nature and each other. Knowledge is understood to be situated, often intuitive, and created through regular, intimate encounters with local surroundings (Worster 1994 [1977]; Sachs 2013). The second version is a scientific-technological project founded on rational, objective, generalizable knowledge that enables the efficient management of nature and people, including maximizing productivity and minimizing humans' negative impacts on the world (Hajer 1995; Symons 2019; Huber 2021).

In short, this chapter argues argue that understanding modernity is crucial to understanding ongoing debates in and beyond political ecology, yet framing debates around being "for" or "against" modernity limits possibilities for creating imaginaries and politics *beyond* modernity.

Of course, any history is necessarily partial, attending more carefully to some moments, places and people and less so to others. Past and present, the most visible versions of debates on modernity within the northern academy and mainstream media have been dominated by white men in the global north. In our review below, we work to show that this frame has global relevance, mindful that it does not encompass the full range of environmental thought globally. Importantly, we focus on modernity and its rejection in order to later move *beyond* modernity and its opposite, to not be limited by this history in how we imagine the future. For this book is rooted in literatures that work beyond modernity and its inverse, premised on the idea that there are other ways to imagine the world we have, and the world we aspire to build.

Arcadians against modernity: Romanticism, postdevelopment, and (some) degrowth

Thoreau's classic, *Walden*, describing the 26 months he spent living in a cabin absent superfluous luxuries, remains one of the fountainheads of environmental thought. His invocation of the pastoral idyll has remained a compelling environmental vision of an ideal life of voluntary simplicity subsumed in the grandeur of a nature that exceeds human comprehension. His scepticism about human faculties of reason and ability to definitively know and exercise dominion over the world prefigures and emblematizes a lineage of environmental thought urging restraint and a conscious reduction in consumption.

While echoing elements of Thoreau's distrust of the hubris of reason and romantic embrace of the land, Césaire's famous lines from *The Notebook of a Return to the Native Land* extend the critique and its countervoice to encompass colonialism and its contestation.

> Eia for those who never invented anything
> for those who never explored anything
> for those who never conquered anything
> but yield, seized, to the essence of all things
> ignorant of surfaces but captivated by the motion of all things
> indifferent to conquering, but playing the game of the world
> truly the eldest sons of the world
> porous to all the breathing of the world.   (Césaire 1939: 37)

In this register, imperial scientific reason is not simply a mode of thought and action that alienates one from the land, but provides an instrumental rationality for colonial dispossession. Thus, the affirmative return to the land is a distinctly political gesture, a decolonial movement by the "native" whose connections to the land had been displaced by imperial regimes of resource extraction and plunder. Against colonialism, there was a revalorization of these historical territorial linkages.

In this section, we outline several components of this broad set of environmental narratives focused on a critique of modern life, modern science, and modern politics. In various registers, there are forms of nostalgic longing for an imagined past prior to modernity or colonialism, and its perceived authentic connection to the earth (Hobsbawm & Ranger 1983). These trajectories of thought, and the communities with which they are associated, defy simple definition. The term "anti-modern" is largely absent from writings, political campaigns and policy prescriptions associated with these environmental movements, arising most in scholarly analyses of environmental thought. "Romantic", the longest-standing term for the cultural counterpoint to modernism, is antiquated and problematic, lacking intelligibility to the contemporary public. "Degrowth" appears to be an emergent umbrella label, yet its intellectual roots span across a range of categories (Demaria *et al.* 2013). In this context, we broadly categorize this anti-modern position as what Worster (1994 [1977]) refers to as "arcadian" and emphasize how arcadian narratives – including some versions of postdevelopment and degrowth – were founded as much through antagonisms with modernity as through affirmations of an alternate future.

In many articulations, the arcadian perspective resonates with what Robbins (2020a) describes as a politics of "less": a politics of orthodox degrowth in which an undifferentiated "humanity" needs to reduce its consumption. Environmentalism has long been characterized as a critique of excess. Contemporary iterations tell us to use less: a standard environmental education tool measures our ecological footprint in additive terms based on the amount of our consumption, and environmentalists urge us to give up luxuries like long hot showers and eating beef (Van den Bergh & Verbruggen 1999; Wiedmann & Barrett 2010). Environmental messaging that stresses the need for "less" often evokes Malthusian graphs in which population growth shoots up, exceeding food supply. While Malthus was no arcadian, many arguing along these lines draw on his thinking to suggest that sustainability requires massive population reductions. Truly sustainable policies are largely seen by advocates of this position as a trade-off with productivity and consumption levels, where real sustainability means a global-scale focus on producing and consuming less. These limits are to be adopted by individuals and shared through cultural change (Hirsch 1976).

Advocacy of less has not always been directly linked to ecological concerns: to varying degrees, implicitly and explicitly, this environmental narrative draws on a long history of *moral* concerns about excess. "While civilisation has been improving our houses", Thoreau (1854: 32) claimed, "it has not equally improved the men who are to inhabit them." Many early romantics were less interested in ecological sustainability than the moral and philosophical search for authentic life, and the curative potential of reconnecting with nature (LaFreniere 1990). Turning away from modern scientific achievements, figures such as Thoreau repeatedly called for a return to the wisdom of the ancients. In a distinct register, the Negritude movement associated with Césaire also harkened back to an ideal of an authentic African heritage, looking to it as the foundation for a postcolonial future (Jones 2010); a generation later such ideas were carried forward in Nyerere's articulation of the politics of ujamaa (Nyerere 1968; Fouéré 2014; Brownell 2020). Kallis's (2019) recent work on "limits" similarly traces the long history of calls for less, positioning degrowth as an extension of a lineage that stretches back to Greek philosophers who advocate for restraint as an ethical practice. The idea that the good life may not be achieved through participation in modernity also underpins a new movement in China, with calls to reject the busy, consumptive narrative deployed by the state (Lin & Gullota 2021).

The entanglement between these moral visions and concerns with ecological processes are often hard to unknot, a point we grapple with further below.

Some bristle at the suggestion that living within limits might be characterized as "austerity", noting that a label of "less" insufficiently characterizes how an arcadian world would be experienced by a global majority. For many decolonial writers, colonialism, as a regime of appropriation and theft, did not endow the colonized with more but less (e.g. Rodney 1972; Frank 1967). Moreover, a number of environmentalists of "less" have explicitly sought to differentiate their project from a politics of scarcity, as signalled by Hickel's playful book title *Less is More* (Hickel 2020; see also Kallis & March 2015; Mehta & Harcourt 2021). They argue that many people need to consume less, but that ultimately their vision is one of healthier, happier lives.

The arcadian view of the good life in the global north took on renewed vitality in conjunction with the mid-twentieth-century birthing of the contemporary environmental movement, which provided even further justification for their critiques of modernity. Many know Rachel Carson primarily as a critic of DDT, and a foundational figure in popularizing environmental concerns through her evocative prose. Yet in *Silent Spring* (1962: 261–2), Carson's criticisms of DDT are rooted in a much broader *social* critique. Characteristically poetic, Carson tells us,

> The "control of nature" is a phrase conceived in arrogance, born of the Neanderthal age of biology and philosophy, when it was supposed that nature exists for the convenience of man. ... It is our alarming misfortune that so primitive a science has armed itself with the most modern and terrible weapons, and that in turning them against the insects it has also turned them against the earth.

The culprits in the story are not only toxic chemicals but the hubris of men. Her gendered language was convention for the time, but some have pointed to Carson as an early ecofeminist, prefiguring later analysis of the masculinism of industrial environmental engineering (e.g. Seager 2003). Feminist environmental thought more generally has called attention to the masculinist approach embodied in our language and practices of exploitation and domination (Merchant 1980; Mies & Shiva 1993; Rocheleau *et al.* 1996). Here, we are told, environmentalists need to rein in not just our consumption or what we produce, but our very efforts to control nature.

Mahatma Gandhi provides another useful example here. His place in the public imaginary is somewhat the inverse of Carson, known mostly as a political figure in the decolonization movement and, for some, an early proponent of alternative development. Yet he argues not only against colonial control, but the broader mentality that rationalizes the domination of people and control of nature (Moolakkattu 2010). His wider concerns resonate with ideas expressed by Césaire, critiquing modernity and the dominant forms of development both in terms of the social and ecological consequences of the colonial logics of dominion. Thus, like Carson, Gandhi observed a rapidly changing world and identified intensifying negative impacts of a modernist vision of economic growth and technological control.

Many Indigenous peoples have similarly drawn upon imagery of a more sustainable and harmonious past to inspire decolonial movements and challenge colonial modernity. Indigenous movements have asserted the need for a decolonial politics beyond the state. For instance, on the centenary of Canadian confederation in 1967, Tsleil-Waututh leader Dan George lamented,

> Oh Canada, I am sad for all the Indian people throughout the land. For I have known you when your forests were mine; when they gave me my meat and my clothing. But in the long hundred years since the white man came, I have seen my freedom disappear like the salmon going mysteriously out to sea. The white man's strange customs, which I could not understand, pressed down upon me until I could no longer breathe.
>
> (George 1967)

As Māori scholar Linda Tuhiwai Smith argues, Indigenous decolonial thought "draws upon a notion of authenticity, of a time before colonization in which we were intact as indigenous peoples. We had absolute authority over our lives; we were born into and lived in a universe which was entirely of our own making" (Smith 1999: 24). Indigenous movements consistently forward their own autonomous frameworks for relating to the land and each other as a counter to colonial orders (Borrows 2010; Simpson 2017; Tomiak et al. 2019). These visions have often explicitly challenged the colonialism underpinning European forms of rationality and reason, strongly defending alternative intellectual frameworks to those of European modernity (Smith 1999; Battiste & Henderson 2000; Archibald 2008; Kimmerer 2013; Kovach 2021). Some authors have stressed the sustainability and communality of Indigenous frameworks for relations, resonating with early decolonial visions in Africa (Coulthard 2014). As we note for many bodies of scholarship, not all Indigenous thought is arcadian, but here we note the importance of some lines of thinking that suggest there is a previous, harmonious and materially simpler world which ought to guide our social and material politics.

Writers like Carson, Césaire, Gandhi and George were ultimately unable to overturn the hegemonic approach to material progress within the twentieth century. After the Second World War and the accompanying formal decolonization across much of the global south, "development" became a new discourse through which to understand socioeconomic change. Modernization lay barely under the surface of this discourse, and at times was its explicit goal (Rostow 1960; Sachs 1992).

Yet modernization, as noted in Box 1, was never really to be. Neither capitalist nor Marxist socialist modernization offers a viable path towards development, and by the 1980s many critical scholars began to note a growing sense of the failures of these two "grand theories" and an "impasse" regarding the idea of development. Further, some even critiqued the very ideas of progress that underpinned development (Birou *et al.* 1977; Escobar 1994; Norgaard 1994), writing in a language that later came to be known as postdevelopment. Instead, some argued, the development discourse that underpins and guides the work of international financial institutions and non-profit organizations remains rooted in colonial modern systems of thought (Rajagopal 2003; Kapoor 2008; de Sousa Santos 2014). Promoting development as a universal ideal, these institutions and their associated enterprises continually export development models from the global north, devoid of recognition of the historical contexts of colonial appropriation and plunder that subsidized northern development trajectories (Frank 1967; Wallerstein 1974; Arrighi 1994; Amin 2009). Moreover, the associated models of development seek to reproduce unsustainable processes of extraction with little regard for ecology or cultural differences. The result has been a consistent pattern of development failures, increased social conflict, worsening environmental crises and continuing northern hegemony (Escobar 1994; Norgaard 1994).

As an alternative to the continuation and intensification of enduring inequalities, postdevelopment theorists have typically argued for pluralizing approaches to social transformation that explicitly rebut the logic of designing or engineering life to fit a singular norm. However, as many have noted, postdevelopment seems to hold together more clearly regarding its object of critique than in its affirmative vision for the future. There has been somewhat of a shift here in recent years, with a more coherent emergent narrative of a pluriverse of local community economies as the root of an alternative political economy (Kothari *et al.* 2019; Acosta & Cajas-Guijarro 2020). While emblematic cases can be found, these alternative paradigms continue to operate largely at the margins, having gained limited traction with existing development policies and frameworks as well as populations targeted for development.

In the global north, a radical environmentalism of "less" seems to be coalescing around the term degrowth as a shared term for an alternative politics (Demaria *et al.* 2013; Kallis 2018; Hickel 2020). Early and some ongoing scholarship on degrowth economics aimed at the quantification of ecological flows and development of calculative tools to manage material limits within a quasi-modernist framework,[1] best exemplified in *The Limits to Growth* (Meadows *et al.* 1972; see Chandler 2018). Yet many scholars working under the broad framework of degrowth emphasize the limits of such tools and predictions, instead embracing "postcolonial, feminist, and anti-racist science and activism [which] have been disrupting conventional politics of knowledge within and beyond academia" (Paulson 2021: 2; see also Nirmal & Rocheleau 2019; Mehta & Harcourt 2021).

Foundationally, the principal concern of degrowth is reorienting the economy and society to a future of less production and consumption. The critiques of modern technology within this literature span opposition to fossil capitalism and nuclear energy to questioning the necessity of kitchen appliances such as the toaster.[2] Some degrowth advocates espouse a discomfort with modern technology and yearn for a rescaling of social life to synchronize with ecological rhythms. Thus, they seek to reorient society from large industrial systems to localized sustainable technologies. This reflects the dream stretching back across the global north and south, from Thoreau in his cabin to Césaire's poetry to Gandhi's spinning wheel, immersing human life within larger ecological relations.

Concerns with distribution were largely absent from early works in degrowth, a point that continues to be noted by critics. Yet later advocates such as Hickel (2020) have made clear that a just distribution requires substantial reductions in the global north while expanding future consumption for many people and many parts of the world. Degrowth advocates have particularly honed their critiques on *destructive* forms of growth or development, while significantly expanding discussion of what a degrowth vision of the future entails. Advocates of degrowth, in a way that resonates with the arguments of Carson, Césaire, Gandhi, and George noted above, suggest that most people would experience degrowth as a safer and more secure world, one in which they would face less exposure to environmental harm.

As evidenced in D'Alisa *et al.*'s (2015) edited volume, some advocates of degrowth draw across romantic, postdevelopment and moral philosophical reasonings to support a wider political ecological strand of analysis and recommendation for the future. The ontologies and knowledge politics that emerge from such intersections challenge easy dichotomies, pointing towards an emergent sensibility neither overly confident about our collective doom nor overly sceptical about the possibilities for shared knowledge. Mapping the unfolding of degrowth thought and its diverse strands is beyond the scope of this book, but for now we note that some seem to embrace arcadian ideas while others are more congruent with the modest approach we develop throughout *Enough!*

The limited success of such movements in garnering widespread political support suggests, however, that historically and for many, they are popularly understood to require sacrifice and concession (Huber 2021). The observation that a smaller economy is not actually what is called for in the majority world (Hickel 2020) makes it difficult to believe in the utility of the term "degrowth" as an analytically appropriate term for a global movement, let alone a banner under which to motivate action in the global south. Further, even those aligned with the wider movement of degrowth have noted weaknesses in developing political theorizations, alliances and analysis (Kallis 2018; Buch-Hansen 2018; see Barca *et al.* 2019).

Thus, through a review of a wider lineage of romantic, pastoral, decolonial, postdevelopment and degrowth environmentalism, we can trace a thread of arguments focused on a critique of many aspects of modernity. The common objects of critique throughout these myriad conversations are excess consumption and scientific hubris, particularly in the presumption that people can and should create regimes of knowledge and power to exercise dominion over the natural world. This line of thinking is founded on a clear and systemic critique of modern society and its ills. Those we have featured are far from the only people we might highlight in a longer history of arcadian, anti-modern environmentalism. Whether one looks to Schumacher's *Small is Beautiful* (2011 [1973]) or the contemporary movement for voluntary simplicity (Rebouças & Soares 2021), these ideas permeate the lineage of environmental thought and continue to inform contemporary environmentalism in its myriad forms.

It is much less clear what those making these critiques propose instead.

What sorts of technologies and practices are in accordance with arcadian views – which would be permissible and how this permission would be written into the world – differ widely across advocates and are often not clearly articulated (Kerschner *et al.* 2018). At times, these critiques are transposed into advocacy for the opposite of the modern – arguments, for instance, to replace masculine hubris with feminine humility, to replace colonial exploitation with Indigenous wisdom, to replace extractivism with localized harmony.

But the inversion of modernist categories traps us within the web of oppositions that modernist thought constructed, creating static and constraining regimes of limits. This is a point that numerous critics of colonial modernity have made, as they express dissatisfaction with the inversion of modernity and the limits of its categories (Spivak 1999; Mignolo 2011).

Thus, there remains a need to articulate positions that exceed the binary of modern and anti-modern, positions that might well accord with the critiques of modernity noted here and – in their affirmative visions for the world – are not defined through binary oppositions.

But we get ahead of ourselves. Before we move *beyond* modernity, let us first dig more deeply into modern environmental thought, highlighting its ambitions, arrogance and limits.

For modern solutions: from imperial conservation to (some) ecosocialism

In contrast to those who stress the limits of modern scientific knowledge, early conservationists celebrated its potential to rationally plan regimes of resource use, enabling sustainable use for future generations. The nineteenth century was characterized by an explosion of modernist approaches to natural resource governance in Europe and its colonies (Grove 1995). Worster (1994 [1977]) calls this approach "imperial" for its associations with efforts to enact dominion over the land in and beyond the colonial era. Wild and unruly forests, for example, were seen as underutilized land, and many argued that private sector deforestation was an inefficient mode of production. The response to both was the same: forests were to be replaced by rationally managed plantations, which were imagined to increase and ensure regular production (Prudham 2007; Braun 2015).

The utilitarianism and arrogance of this way of thinking became the subject of much critique by the emergent mainstream environmental movement of the 1960s and 1970s. Yet what it meant to be "environmentalist" became increasingly ambiguous and open-ended as modern solutions again became integrated into the project of environmentalism. In the global south, postcolonial countries challenged the environmentalism of limits as a cultural export from the former imperial metropole that called for enduring austerity in the former colonies, a concern famously vocalized by India's Prime Minister Indira Gandhi at the 1972 Stockholm Conference (Gandhi 1972; Rangarajan 2006). Rather than rejecting the project of environmentalism or development *carte blanche*, they blended desires into the increasingly popularized concept of sustainable development (World Commission on Environment and Development 1987; Redclift 1992). For many, this meant the broad trajectory of modernization could remain, but with a light touch of consideration for environmental impacts.

More recent work draws on this longer history of efforts to know and control nature through imperial science under the term "ecological modernization" (Hajer 1995; Symons 2019). This form of environmentalism is framed as an alternative to the technological pessimism of the environmental movement of the 1960s and 70s, rejecting its story of "less". It is also proposed as a more specific programme than the ambiguous politics advanced under the vague term "sustainable development". Thus, since the 1980s, modernist thinking has again become folded into particular strands of environmentalism, reformulated as *part of,* rather than the *object of critique of,* environmentalism. Ecomodernists challenged the anti-modern politics espoused through the 1960s and 1970s by figures such as Carson. Simultaneously, corporations, seeking to limit their risk exposure and the costs associated with environmental campaigns and organized boycotts, appropriated the discourse of ecological modernization, adopting policies of purported corporate social responsibility and sustainability. To an extent, this reflected and reinforced a shift within environmental organizations towards increasing pragmatism. Rather than challenging foundational political and economic structures, numerous mainstream environmental organizations focused on articulating policy remedies that better accorded with existing power relations (Hajer 1995). Foundational to this line of thinking is that environmental sustainability can happen with economic growth and increasing consumption.

The political economy of ecomodernism remains the subject of important debate, and we return to such questions in subsequent chapters. In its most well-known contemporary form, ecomodernists often work within the political economic ideas of capitalism, urging the harnessing of science and progress towards a greener future, as investors shift to support experimentation in environmental solutions (Goldstein 2018). However, many ecomodernist activists adopt a more interventionist stance, emphasizing the important role of the state in regulating capitalism to moderate its tendencies to externalize environmental costs onto the public (Symons 2019). Both of these positions share an optimism regarding the possibilities for states and markets to fuel investment and technological innovation, allowing both improved sustainability and continued increases in material well-being globally. Mann and Wainwright (2018) provide a provocative political ecological analysis of what such a future might entail: a state empowered to redress the climate crisis in a capitalist world may well produce a Climate Leviathan, boldly enforcing change and inequality.

Others, however, have pointed towards a different vision of ecomodernism, one that embraces modernity but not its efforts to reconcile environmentalism within a capitalist political and economic order. It distances itself from the limits of the "real world socialisms" of the twentieth century and works towards a more democratic version of ecosocialist modernity.

Specifically, modernist ecosocialists have argued that the problem with ongoing ecomodernist politics is not modernity writ large, but the ways in which capitalism has captured and directs science and technology.[3] Many green socialists thus argue that a *capitalist* version of ecomodernism is inherently unsustainable (O'Connor 1991; Kovel 2002). The solution, from an ecosocialist modernist perspective, is not to eschew *modernism* but *capitalism* and its broader logics of accumulation (Huber 2021). This is explicit in Green Marxist scholarship, which draws on a longer history of Marxist theorizing to suggest that removing scientific practice from the direction of capital will open wider horizons for technological progress and innovation. A strong, controlling state is justified to make radical interventions given the scale of crises (Li 2009; Davis 2010; Bigger 2012). Mann and Wainwright (2018) also consider what one version of this might look like for climate change, labelling it "Climate Mao", in which a strong socialist state creates rules and enforces reductions in carbon emissions.

Ecosocialist modernism, then, is premised on a modernist belief in our capacity to appropriately know and control nature and the possibility of embracing modernity's technological promise. Advocates implicitly or explicitly believe in the possibility of separating what many concerned with colonialism reviewed above saw and see as linked: the domination of nature and the urge to know, control and govern populations (see Collard *et al.* 2016).

We (Mary and Tyler) believe, as environmental critiques of modernism highlighted above, there are inherent dangers in placing unwavering faith in science and our capacity to rationally manage complex ecosystems. The broad application of policies of scientific management to public lands extended the project of settler colonialism, displacing Indigenous systems of land management (Spence 2000; Kosek 2006). Moreover, management regimes that sought to maximize yields – for instance, suppressing fires in forests and eradicating rabbits from rangelands – disrupted ecological regimes and led to the long-term degradation of those environments (Earley 2004; Sayre 2017). We are also sceptical of the possibility of drawing a clean line between the natural and social here, of retaining a belief in controlling nature and letting go of efforts to control the social. For the long history of social theory shows the entwining of social and ecological sciences, with spill-overs of ideas and practices, consciously or not (Worster 1994 [1979]). Further, much of the work of ecologists and climatologists charting biophysical dynamics has not given us complete models but rather employed science strategically to highlight the dangers of an unbridled faith in technological advance and unlimited progress. As noted in Box 1, and elaborated on further in Box 2, decolonial and other scholars have insisted that modernity is a totalizing imaginary: we cannot pick and choose, separate its good and bad. We need a different imaginary, a new story of what can be.

Box 2 The lie of modernity

Let us, Mary and Tyler, not be ambivalent regarding our position on modernity: a just and sustainable modernity is a beautiful dream-nightmare, a dream-nightmare that underpins many of our cities, our politics and our economies.

In the wake of decolonization, in the middle of the Cold War, the new language of "development" led many to believe that someday the whole world might be modern. Countries who needed to develop were given economic and political plans, funding and expertise and told that they, too, could become modern.

If only.

If only they followed the right steps, toed the lines, made the right sacrifices.

Yes, there was racism and poverty "there" in the undeveloped world, and yes, it was "here" in already modern places: but these were fixable! We-the-developed-world did not need to share what we had – if we did, "they" would not have earned and appreciate it. "We" needed to teach "them" to work, to build, to sacrifice and make a world like ours.

And many in the global south agreed: we-the-undeveloped-world need to work, to sacrifice, and thereby build and deserve our own modernity.

Who would doubt such a story?

At various moments, we all could look at the global north, where cities and politics and economies were beautiful (nightmares). Imperfect, yes! But headed, quickly, in the right direction. In the nineteenth century, slavery was ending! Voting rights were expanding! Industry was making amazing gadgets! In the twentieth century, homes were being built in the suburbs with running water, with such abundance that manicured lawns were grown in the desert! Roads were built, and cars filled the suburban garages! Energy was abundant, and we were on our way to harnessing nuclear power to make it even more so!

Many of us (including Mary and Tyler), particularly white children with economic privilege in the global north, grew up being told this was our story, this was our legacy: our ancestors had worked hard, sacrificed much and were imperfect, but what a world they had built, and here it was, ours to enjoy! We-in-the-global-south were also told this was our story: we would be the generation who would work hard, sacrifice much, and build a modern world for our children!

So many foundational cultural beliefs were forged in this story. So many of us believe the world is plentiful and there is opportunity and resources (land for the tilling) for any willing to work. We also believe our material desires are insatiable, and therefore that there will never quite be enough, so we must all work hard to fulfil as many wants as we can.

Equality, property, markets, education and rights will be our guides, we will work hard and enjoy the fruits of our labour!

This is a broken story.

A story many still believe, and many more (including for a long time, we the authors) still *want* to believe. It is a story that justifies the world we have, the sacrifices and inequalities and morality that underpins it.

This story calls on us to act, but it does not question the direction in which we are told our world is headed.

We (collective) were told that the good, modern future would be built through controlling nature and rationally creating good citizens.

But here is the threefold lie: the illusion of modernity – the cities and politics and economies we have, north and south – were built through, and rely on, extraction, exploitation and coercion. They require the taking of material resources from the global south through colonialism, the creation of fraught material flows from the rural to the urban, as well as ongoing neocolonial, kyriarchal, globalized relations. They require the exploitation of nature, most emblematically the use of fossil fuels, but also the diversion of water, the pollution of land and sea, flora and fauna. They require underpaid labour, often coerced and rarely done in conditions where there are other viable livelihood possibilities.

But nature cannot be controlled. Citizens rebel against stifling uniformity. Modernity never was and never will be.

Facing these truths is difficult. Facing this history is deeply uncomfortable. Important.

Necessary.

But also, not enough.

This story of the past matters for helping us (collective) to see why we cannot carry on as we have been, why inequalities cannot be patched over through doing more of the same.

It is not enough to let a dream die.

We need a new story, a story that is less ephemeral, more rooted in the world we have.

We need modest aspirations, not modernist dreams.

So what? Now what?

What are we to do with these critiques and concerns with modernity and its inverse? Can we believe in modernity's promises without enrolling its moral failings? Can we draw on rational, calculative science without humanity being subsumed to algorithms? Can we fight for justice without a universalizing notion of humanity?

Might we begin to make sense of the world we want not as an eclectic or partial adoption of a revised modernity but, possibly, as something else?

These are the questions that we (Mary and Tyler) sat with for many years. As people concerned with environmentalism and development, and whose research has been situated in peripheralized places, ideas and literatures, we struggled to find our way out of this seeming impasse. Our doubts kept us from being wholehearted activists for either side of the environmental divide or the postdevelopment debate, for the solutions proposed seemed inadequate, detached from the worlds in which we dwelled and likely to generate other harms.

We wanted to know what we believe *in*, what we could stand behind, what we were for (see Introduction; Ferguson 2010).

So, as good academics, we (Mary and Tyler) read more beyond our core literatures. As good dwellers of peripheralized places, ideas and literatures, we searched beyond the canon. And in the next chapter, we point towards some of the works that inspired us in our search for ideas beyond the modern/anti-modern impasse. We then sketch what it might mean to work beyond this binary by outlining a modest imaginary and a modest approach to politics.

2

NEITHER MORE NOR LESS: CULTIVATING
A MODEST POLITICAL ECOLOGY

Flora lives in an informal settlement in Cape Town. Most days, she travels from the city's edge to the formerly white, but no longer legally segregated, suburbs. There, she cleans houses (and yes, apartheid has ended but all her employers are white people). Flora earns enough money that she does not worry about rent, food or clothing most months, and is saving funds to ensure her daughter gets a tertiary education, likely at a technical training college. While she is at work, she uses the toilet, a toilet that flushes the waste away through the sewer to the city's edge. At her own home, there is a shared tap where, usually, she can get water for drinking and washing. There is no sewer. There is no toilet. There is a shared facility a short walk away, one that is sort of run by the city and sort of run by the various community members who have taken an interest in the toilets. And by "taken", well, we do mean taken; and by "interests", well, we do mean that these neighbours take money from the city and users to ensure the toilets are usually clean enough, usable enough.

Flora knows that faecal matter is something to avoid, that it is dirty, unhygienic, unsafe: as is true for most people, "education" is not the problem here. She knows that in the rural area where she grew up, no one worried about sanitation: there was a place you would "go" and the water from the streams and rains would take it away. In the city, she is concerned about sanitation, but it is far from the most important of her worries. She knows that modern sanitation (a flushing toilet connected to a city-wide sewer) is a long way away from her home. It is in the suburbs where she works. It is in the promises of some politicians and activists who insist that a flushing toilet will arrive sometime in the future.

Flora doubts that modern sanitation is for her, in her home, in her lifetime. Maybe, at one point, thirty years ago when the country was full of post-apartheid rainbow nation hope, she believed it might happen. She sees the excesses of water on the lawns of the green, formerly white suburbs, but lately there have been droughts, water is increasingly expensive and scarce, and even the lawns of the homes she cleans are becoming a little less green.

Flora's story resonates with billions of urban residents across the global south, and increasingly some in the north too. The modern imaginary of infrastructure – including the idea that everyone was going to someday get a flushing toilet – is increasingly doubted by ordinary people and experts alike (van Vliet *et al.* 2010; Sutherland *et al.* 2014). There are plenty of technological alternatives, some of which work better in some places, under some arrangements, than others. How are we to understand these alternatives, which are acceptable and which are not? How might we move back and forth between toilets and a wider understanding of how the world works, what is just and what is sustainable?

Politically, it is easy to insist that "enough for all" means "modern sanitation for all". Ecologically, it is easy to insist that no one should use modern sanitation systems, for they use water at unsustainable rates and concentrate waste in environmentally detrimental ways. A modest approach to the political ecological goal of enough for all means insisting that just and sustainable sanitation for all is possible, and works to think through how it can be achieved in a world that is uncertain.

As we can clearly see in the most mundane of examples, our options for the future are not confined to pursuing modernity (the unsustainable flushing toilets of the wealthy suburbs) or describing our pursuits through their antagonistic relationship with the modern (low-density defecation washed away and sustainably integrated into nature's nutrient cycles). We (Mary and Tyler) believe that people can, and ought to, reject modernity without adopting its inverse. There is no shortage of empirical examples that defy this easy binary: we could list dozens of designs for toilets that are neither modern nor arcadian.

In the previous chapter (as well as in our wider writing, see Lawhon *et al.* 2023a; Gergan & McCreary 2022), we have shown ways in which modernity remains a useful anchor in debates about justice and sustainability. Thinking about modernity is *analytically* useful for understanding prevailing imaginaries. Yet this book is premised on the idea that this is not the only way to understand the world. The conceptual topography beyond these established reference points is less well established (see Box 3), but we are not without guides: this is not uncharted territory.

Across and beyond the natural and social sciences, in and beyond the academy, people are and always have been living, experimenting and theorizing outside the modern/anti-modern binary. In myriad discussions, scientists, scholars, activists, artists and ordinary people are asking what can be known and how this should inform the ways that people interact with the environment and each other. They are grappling with uncertainty without letting it stop them from intervening in the world, and believing it can be better.

In the first half of this chapter, we briefly point towards cognate and inspirational ideas, emphasizing literatures that centre uncertainty and plurality in social and ecological theory. What we attempt here is far from comprehensive: many others have provided more detailed engagements with these ontological inquiries.

Box 3 This Changes Everything

This Changes Everything is the title of a book by Naomi Klein (2014) and documentary inspired by the book (Lewis 2015). Mary uses the film in her teaching, and it has become a touchstone for her thinking over the last few years, thinking reflected in *Enough!*

The book is subtitled "Capitalism vs the Climate" and spends a considerable amount of time reviewing common anti-capitalist arguments. It also describes numerous cases of what Klein calls "blockadia" – transnational conflict zones where people are actively opposing extractive capitalist development, often led by Indigenous communities, such as those that Tyler studies.

There is also a second line of argument in the book, one that comes after the anti-capitalist critique (and is more central in the film). This is a concern with what Klein calls "extractivism". The critique of extractivism parallels longstanding concerns with modernity in environmentalism. Here, the focus is on challenging a dominant "cultural story", a story in which our collective belief that we can control nature underpins our engagements with the world.

How do these two points – a critique of capitalism and a critique of modernity – work together? This is not entirely clear in the text or film. Some have suggested a tension or conflation here, urging Klein and her readers back towards more orthodox concerns with capital and retaining the possibility for just green industrialism (Huber 2018). We disagree.

Huber suggests that Klein relies on classic twentieth-century environmental ideas for her solutions, mixing respect for Indigenous ways of life with reconnecting with nature. There is, we agree, a little of this here and there.

More centrally, however, we see *This Changes Everything* as a provocation towards thinking about what comes after a critique of capitalism and a critique of modernity – the fundamental questions that we (Mary and Tyler) aim to address. Both book and film point to people saying "no" to capitalism and the belief that we can control nature. But they also gesture to a growing interest in prefiguring a "yes" that is not clearly arcadian. Klein and Lewis seek to understand not just what we are against (fossil fuel capitalism) but what we are *for*.

The "no" is clear, but the "yes" is less so. *This Changes Everything* develops much less analysis of what exactly it is about particular cases that makes us able to see them as part of the "yes". Crucially, however, the "yes" that we are shown is not a romantic return to a pre-industrial past (see Chapter 1). Huber (2018) creates a strawperson here, one easy to cast aside.

What, exactly, constitutes this "yes" in and beyond *This Changes Everything* is, however, somewhat unclear. Mann and Wainwright (2018) provide some useful guidance, drawing in part on Klein's work to consider a non-capitalist future with distributed sovereignty. In contrast to the Climate Leviathan and Climate Mao noted in Chapter 1, they call this "Climate X".

Enough! is written to build on the analysis and narrative of *This Changes Everything* (one of many sources of inspiration) and pushes forward towards a somewhat sharper vision of what "Climate X" might entail and how it might be enacted. There is a problematic cultural story, a story that has been shaped by – but ought not analytically be subsumed within analyses of – capitalism. We need a new story, a story that connects to real world cases but also has a clear narrative, one that explains *why* some examples are part of the story of "yes". We need help knowing what we might want to lean on, add to, learn from. *Enough!* builds on Klein's important work, and we hope might be read as a next step in this wider project, telling a story that places both the "no" and the "yes" into longer intellectual genealogies, and works to more clearly identify and explain the "yes".

We also work to think through how to build the "yes". Klein (2014) does mention basic income as part of her vision, but this point seems supplemental, something nice to have but not interlinked with her argument. As we show throughout *Enough!*, we agree with Klein that basic income ought to be part of the story of "yes" – and work to tease out the ways in which it can be integral to, and may well enable, worlds we want to say "yes" to.

Throughout *Enough!* we hope to make clear our intellectual debts and the resonance of our argument with many allies. Our intention in reviewing key influences here is to demonstrate that the wider modest imaginary we name has diffuse roots and accords with the cutting edge of much social, ecological and socioecological thought. We work not to advance conversations on ontology but instead to tease out the *political* implications of these insights. Our focus, thus, is ontopolitics – considering the implications of debates over "what is" on "what can and ought to be". Specifically, we seek to explain how particular lines of social and ecological scholarship shape our approach to a politics of enough.

The second half of this chapter offers a name for our approach to politics: modest. We note our initial misgivings about putting new vocabulary into a crowded academic lexicon, yet ultimately demonstrate the need for a new term, and why "modest" is an imperfect but, we hope, useful analytic. We then explain what we call a modest imaginary and point to three key components: emplaced sustainability, entangled autonomy and anticipatory guiding. This chapter, thus, provides a framework that guides the remainder of the book, in which we develop a modest approach to the political ecological goal of enough for all.

Uncertainty and indeterminacy in socioecological thought

Modern approaches to science were founded on the idea that hypothesis testing of closed laboratory experiences or controlled studies could provide explanations for the workings of the world (see Box 1; Chapter 1). These narrow and reductionist paradigms were never widely adopted by naturalists, making ecology an important field of study through which to understand changing ideas about the possibilities of science. Many sought to understand the complex networks of ecological relationship between flora and fauna, convinced that ecology could not be read off from a single organism or set of controlled variables (Golley 1993; Wulf 2015). Efforts to statically capture these webs of relationships were consistently challenged by their complexity (Raby 2017).

In this context, let us reiterate: non-modern ways of understanding the world have always been part of natural sciences! We are not advocating for a rejection of science, but instead a modest approach to it.

As ecology emerged as a science that sought to explain the logic of nature, it did so based on theories of complexity rather than reductionist laws (Worster 1994 [1977]). These efforts have inspired some versions of resilience thinking as well as climate science: there is an acceptance that relationships cannot be distilled to causal linkages and summed up in a mechanistic way. Instead, complexity is to be modelled, approximated, estimated in ongoing and iterative ways (Edwards 2010; Kitchin 2014). Our ever-increasing capacities for computing enables modelling of more and more factors, yet, despite longstanding efforts and enormous public funding, science continues to be unable to consistently predict what will be (Edwards 2010; Sobel 2014). Instead of definitive answers of current and future environmental states, biophysical models of oceanography, climatology, hydrology or ecology suggest *probabilities* within complex, dynamic systems.

In this way of understanding, nature, due to its immense complexity and unpredictability, continues to – and always will – surprise us. In this way of understanding, science still has a place: this is no anti-modern *rejection* of the utility of data, observation or knowledge. It is an acceptance that uncertainty is not temporary, that the world is fundamentally indeterminate, and thus our knowledge of it is partial and imperfect.

How are we to understand the assumptions and implications of this science for justice and sustainability?

Some have doubled down on the drive to calculate and control, to limit uncertainty in the hopes of eventually reaching sufficiently full knowledge (see Box 1), largely retaining modernist hopes. But this is not the only way in which people have responded to uncertainty.

Of interest to us here is a line of thinking by those in and beyond the academy that accepts uncertainty as something that cannot be fixed: it is a fundamental condition of being in a changing world (Grosz 2011; Haraway 2016; Tsing 2015) and must shape how we collectively think about environmentalism, governance and politics (Mehta *et al.* 2001). This perspective is not arcadian: there is no harmony which can be found with an always changing and unknowable nature, but there are patterns in our world that can be observed and understood. Understanding, then, is not complete but recursive and partial. The patterns that become visible are not deterministic: they cannot tell us the future. They change for intelligible and unintelligible reasons.

Modernist efforts to plan, manage and control must, then, be replaced by efforts to experiment, anticipate, learn and guide. Prediction is less useful, and more dangerous, than practical wisdom or phronesis (Flyvbjerg 2005). Understanding is best viewed as an interplay between the explicit information that we have catalogued and the tacit knowledge and attunements that may not rise to consciousness. This understanding is always generated in place, but is not site specific: practical, engaged and tacit wisdom can be best enabled by shared learning across local experiential knowledge *and* expertise derived analogically from insights produced across multiple cases. The kinds of knowledge produced through such interactions are not abstract and replicable, but situated and practical. They are not defined by a search for universal truth and explanation, but by a search for effectiveness and the relations that enable that effect (Ali 2007; Vasudevan & Novoa 2022).

This way of thinking is most evident in fields such as ecology, planning and design studies, fields motivated by a desire to use knowledge to intervene in the world. They share a recognition of the limits of conventional laboratory methods and a search for ways to intervene despite uncertainty.

What does this mean for political ecology? Many of us (political ecologists) have comfortably used the unsettling of modernist science to critique the hubris of modernist biophysical sciences, pointing to the limits of efforts to confidently know and control. We have even critically studied uncertainty, considering how it is framed and navigated, providing analysis and critique of the politics of uncertainty (e.g. Eren 2017; Mehta *et al.* 2022).

Such unsettlings of science also have implications for how we think about our role as political ecologists, implications that recent works have begun to tease out. For political ecology, like many other contemporary academic fields, continues to bear the marks of its foundational frameworks: Marxism and poststructuralism. Both are rooted in quite different approaches to knowledge, science and scholarship, ones that have enabled radical critique. Yet, as many have pointed out, the relationship between political ecology and a broadly writ "policy/practice" has long been troubled and remains the subject of much debate (Walker 2007; Rocheleau 2008; Heynen & Van Sant 2015).

Most political ecologists agree that the purpose of scholarship is to shape our world (see Castree *et al.* 2010). While political ecology has long been an engaged science, it has often been noted to be better at analysis and critique than action. Both Marxism and poststructuralism give political ecologists strong frameworks through which to assess the problems with the world we have. Marxism has long made clear the world "we" want, but in practice Marxist political ecologists write much more about the problems of neoliberalism than the pathways and worlds beyond it (e.g. Heynen *et al.* 2007; see Ferguson 2010). Some scholars have reflected on what our role ought to be, and the extent to which we ought to critique the imperfections of our allies (Goldfischer *et al.* 2020; Desvallées *et al.* 2022). Other texts, often rooted in postcolonial critique and decolonial thought, have pointed to examples of more progressive socioecologies (Daigle 2019; Heynen & Ybarra 2021; although see McCarthy 2002 on the analytical ambivalence of "what we are for" in political ecology), at times centring the role of scholars as "giving voice" rather than adding analytical value.

Political ecology and political ecologists' relationship to the future, however, has been more tenuous, more ambiguous and ambivalent. Critique tells us what is wrong, and let us (Mary and Tyler) be clear we are fully supportive of critique. But it does not help us know what we are for (Ferguson 2010), nor how to build towards better worlds. Many political ecologists, ourselves included, are wary of experimental and interventionist methodologies, wary of telling others what the answers are. We dwell in the long history in which academics did – and continue to – intervene in problematic ways. From defences of racism and colonialism to providing techniques for extraction and contamination, the academy has a deeply fraught relationship with practice. We do not in any way want to minimize this concern.

But as we argue throughout *Enough!*, it is possible to work beyond these options.

A modest approach to scholarship means taking seriously the possibility that we as scholars have something useful to contribute and finding ways to engage that go beyond critique or pointing uncritically towards our allies without placing ourselves as all-knowing arbiters of truth(s).

Our proposal here, then, is rooted not in the question of more or less critique and engagement. It instead is rooted in an emergent body of political ecological scholarship that takes a more tentative position regarding what it means to know, including the utility and validity of the tradition of radical scholarly critique[1]

(Braun 2015). For if knowledge is no longer a thing to "have", it can no longer be (and never was) "ours" to wield from an ivory tower. In this context, we (political ecologists) must think through how we know what we are for, and how we imagine changing our world. Doing so means doing our science differently, less confidentially, more reflexively, even experimentally, and we consider *Enough!* as part of a wider move of opening political ecology towards this orientation (Leach *et al.* 2007; Braun 2015; Harris 2021). Critique remains part of our wider toolbox as proposals, prefigurations and experiments are undertaken reflexively, and analysed within wider configurations of power. Yet the grounds on which it is made, and its centrality to political ecology, are increasingly suspect.

There is not a definite and singular answer to what this means for the role of science and scholarship, but we draw here on recent books by Escobar (2018) and Grove (2018) to demonstrate possibilities. Both use the language of design and review changes in this field, showing a move away from the confident application of modernist knowledge towards more experimental and uncertain engagements (see also White 2020). Escobar (2018) puts design in conversation with postcolonial and postdevelopment thought, showing the potential utility for thinking across these fields. What emerges is both a postcolonial/postdevelopment critique of design *and* a set of contemplations about how a more open process of design might inform the alternate world-making demanded by many in and beyond the global south. This way of thinking informs the modest approach we outline below.

Grove's (2018) book on resilience forms a productive parallel, again pushing political ecologists to engage with questions of the future from a more capacious and uncertain position. While many in and beyond political ecology suggest that resilience thinking is simply a new version of modern, positivist thought – and indeed, there are versions of both design and resilience that are modern – Grove traces its conceptual origins, showing its roots within a deeply not-modern, practice-based, designerly paradigm. This is not to say that there are not real-world uses of the term "resilience" that operate otherwise: examples proliferate in which modern efforts to know and control are largely retained under a new and shiny title, and these have often been the object of critical analysis (for example, showing that resilience is a new package for old techniques of neoliberal governance, Nelson 2014; Jones & McCreary 2022).

Yet, Grove (2018) shows that understanding resilience genealogically enables political ecologists to go beyond critiquing a new label for old modernist practices, and see how resilience thinking can be used to open new ways of understanding – and changing – our world. For, in drawing on designerly thinking and its understanding of systems and complexity rather than a modernist empirical science in search of objective truths, it provides a very different understanding of "the relationship between truth and control" (2018: 3). Read through a longer history of knowledge, resilience thinking is part of the growing body of academic work that lies outside the modern/anti-modern binary: it rejects the possibility of a modernist relationship between scientific truths and control without rejecting the possibility that knowledge can contribute to human welfare. It changes knowledge from an abstraction to a context-based insight that can only come from experience and interaction with the world. The explanations that result from such engagements are necessarily inchoate, not only partial but also in a constant process of reformulation in a changing world. Moreover, they will always need to be reworked to make sense elsewhere.

Resilience thinking, in its orthodox version, is about how to maintain stasis in the face of change. Writing around the Anthropocene has expanded this to think through life amid contemporary ruins. Yet Grove suggests, and Wakefield (2020) develops in her work, there is underexamined potential to draw on resilience thinking to better understand *transformation.* In other words, we might use resilience thinking to understand why unjust, unsustainable worlds are difficult to change, and use designerly thinking to help find a way out and construct new worlds (see also Scoones *et al.* 2015).

In this sense, there is some affinity between resilience thinking for transformation and some writing in the field of sustainability transitions: the question shifts from understanding how to become resilient towards how to change in order to become more able to live in an uncertain world (Walker & Shove 2007; Collier *et al.* 2013; Zanotti *et al.* 2020).

And it is within this wider literature that we position our arguments for a modest political ecology of enough: we are searching for what and how to change in order to live well amid ecological uncertainty.

Plurality and uncertainty regarding justice and the good life

The unravelling of modernity has not only had implications in ecological and socioecological thought: the social sciences and humanities too have also been characterized by the fracturing of a universal and singular understanding of "justice" and what this means for a shared or universal understanding of what it means to flourish. Here we briefly review key arguments, and provide more detail of these arguments and literatures in the chapters that follow.

Most (but again, not all) Western social theory has historically been underpinned by a modern imaginary, premised on a knowable, rational human, or for some, a humanity composed of multiple, distinct human populations that could be scientifically distinguished. Sociology and – generations later – development theory were founded on modern stagial thinking and teleology (see Chapter 1 and Box 1). The very categories on which such studies are premised are increasingly subject to critique, calling forth anti-essentialist identities that work against the knowability of humans (including but not only through their demographics) and universalized, singular notions of the good life. Further, the idea of a clear pathway for civilization and a known outcome has largely been rejected in and beyond sociology and development studies. For some, the loss of such a singular vision resulted in postmodern relativism or the development impasse, both characterized by a sense that progress might well be unknowable and impossible (Harvey 1990; Jameson 1991).

There are, however, many moving beyond this impasse, beyond this relativism and towards a more affirmative, open-ended vision of justice. Anti-essentialism here is not a *rejection* of identity as central to the sense of self but a rejection of its knowability, nameability and predictability. What justice means is complicated by such a narrative, for needs can no longer be defined for humanity, nor for "women", "white people" or "the poor" (or even a crassly intersectional approach to these, see Box 1) and then met through codified processes. Instead, justice is understood in terms of freedom, self-expression, autonomy and radical democracy (although, as we show throughout the chapters that follow, our conception of these terms is distinct from liberal humanist versions) (Laclau & Mouffe 1985; Kelley 2002; Hardt & Negri 2009; Davis 2015; Gilmore 2022). Similarly, there are ways to imagine better worlds that do not rely on a universalized notion of improvement. Letting go of a linear narrative of development opens interpretations of what is meant by betterment, progress and development to as yet unknown possibilities.

In this context, we also need to rethink the assumption that our material desires are boundless, and that if given a choice between autonomy, time and consumption, we will optimize for the latter. Countless studies have shown that, above a certain threshold, there is no correlation between income and happiness or life satisfaction (Kahneman *et al.* 2006). As we turn to in Chapter 5, there is a growing scepticism that a compulsion towards unbounded accumulation is somehow inherent to the human condition. Yes, many of us seek to not have less than the norm; the arguments we develop in *Enough!* are largely in keeping with the idea that our material desires are strongly shaped by our social context. Yes, many studies have shown that some people are motivated to have more than their peers. Such findings tell us much about the values and experiences of research participants, but overrepresent subjectivities shaped by modern capitalist contexts (Henrich *et al.* 2005). The arguments we develop in *Enough!* draw on recent scholarship that insists on a broader geographical and historical analysis, pushing back against the naturalization of competition, individualism and overconsumption. Instead, viewing subjectivities as shaped by context, we consider the kinds of subjectivities that might be cultivated in a more secure world, one in which everyone was assured enough (see Interlude and Chapter 6). We do not believe such a world would somehow end comparison and competition – nor ought it. But we do think that it might well reorient our relationship to consumption, each other and how we spend our time.

It is within this wider literature that we position our arguments for a modest political ecology of enough: we are searching for what and how to change in order to live well amid plural notions of what the good life entails.

This whistlestop account of uncertainty – of our inability to know and predict nature or each other – is far from a comprehensive account of contemporary social, ecological and socioecological theory. Key to the argument that follows is, in short, that we write in alignment with much contemporary thought that insists on moving beyond modernity and its inverse, embracing the possibilities of partial knowledge and the impossibility of control, and a role for scholars in shaping (not knowing but not only critiquing) what comes. Importantly, alternatives to modernity have always existed in and beyond the Western canon, in and beyond the academy. What is different now is not their presence, but the growing acceptance of their global relevance.

There are, of course, risks and difficulties in working across social, ecological and socioecological theories, and our intention here is not to suggest that we converge diverse accounts into a singular narrative. We ought not equate the uncertainties of ecology with uncertainties about what justice and flourishing are and how they ought to be pursued. There is, however, a synergy here that seems impossible to ignore, a crossover that Worster (1994 [1977]; cf. Charbonnier 2021) shows has long existed between ever-changing ideas of society and ecology. We see potential in a shared rejection of modernity and a search for something that goes beyond it, a bubbling sense of potential for a new, emergent ontopolitics, an ontopolitics of complexity, uncertainty and possibility (Wakefield 2020).

A modest imaginary of the world that is

In this section, we present an initial outline of a modest approach, an approach that is deeply indebted to many other scholars and literatures for shaping our emergent ideas.

In working through the literatures noted above over the past several years, we read with much hope that we might find a term that captured the wider framing that we sought to work within. There are, no doubt, plenty of terms that have been developed in the past forty years to speak towards this rejection of modernity (see Box 1). Yet we struggled with existing vocabulary, wondering how we might stretch and repurpose terms with similar usages and complex histories and constituencies. So many terms we encountered challenged modernity and its various allies by positioning themselves in reference to what they are not: postdevelopment, degrowth, postmodern, postcolonial, postnormal, anti-essentialist. Each of these points towards slightly different sources of concern and possibility. All are largely oriented towards a critique of modernity and its various iterations, components and inflections in particular fields of interest (ecological, sustainability, development, race and gender studies).

Our ambition here, instead, is to develop a framing that is distinct from, but not defined through, its opposition with modernity.

We also considered various additive phrases, wondering if we might name multiple allies. Yet we did not want to imply that our ideas synthesized every-thing about, for example, feminism, resilience or decolonial thought under a single umbrella, as we do not intend to reduce or subsume these different ideas within one overarching term. For some feminist thought remains rooted in modernity; some use resilience and designerly thinking in neoliberal ways; some postcolonial politics continue to embrace modernity. These terms have been used, rejected, debated, repurposed, claimed and reclaimed by a variety of interests. Instead, we aim to draw on and tease out a thread shared by many, without presuming it is shared by all.

In the end, we decided not to deploy an existing conceptual term or combin-ation of them towards new, narrower or more specific ends. Instead, with some trepidation, we embrace a term that exists and work to fill it with conceptual, ontopolitical weight. We hope the reader will forgive the hubris of a new term in a crowded academic lexicon, and see our effort here as but one moment in a longer collective process of charting the contours of this imaginary.

In what follows and in the remainder of the book, we work with the idea of a modest imaginary, turning it into a modest approach through which to pursue enough for all through the creation of modest economies, states and livelihoods.

A what and why of "modest"

We propose modest not as a middle ground between modern and anti-modern but as an alternative that collapses binaries produced through modernity.

We also like the phonetic similarities of modern and modest, a useful way to signal that modest differently occupies a similar conceptual and linguistic space.

In its common usage, modesty is a moderate or medium amount of some-thing, but there is more to the difference between more, less and modest than quantity. The word modest is used not so much to describe an amount, but a *quality*. Etymologically, it is derived from a combination of the French term *modeste*, which connotes the quality of being "free from excess", as well as the Latin *modestus*, which associates the term with further moral qualities as something restrained, temperate, well-behaved, seemly, decorous, unassuming.

We embrace some parts of the vernacular meanings and etymology more than others as we add conceptual weight to this term. To be unassuming is to be free of hubris, a point that resonates with the ongoingness of uncertainty described above. Being free from excess is important, and in Chapter 5 we elaborate on the freedoms that come from ending a pursuit of excess; this is quite different from asserting the morality of limits and simplicity described in Chapter 1 (see Kallis 2018). We distance ourselves from its association with being "well-behaved" or "seemly", and the implication that modest must be quiet and demure.

Let us say loudly: we name this imaginary and politics "modest", but ours is not a quiet nor well-behaved approach.

Ours is not a vision of small dreams or acquiescence to the existing inequalities of the world. There is nothing quiescent about a modest approach. Modest is radical in its effort to displace the ongoing remnants of modernity, to rein in the excesses and displace the amorality of postmodernity.

Modest, as we deploy the term, differs from hubris and humility in crucial ways. Unlike the hubris of both the modern and postmodern, modest is not fixated on superficial, spectacular or external appearances. It is an affirmation of the centrality of mundane, everyday engagements with possibilities and constraints. Humility often implies unimportance, subservience and submission. While many use modest and humble in similar ways, our intention here is to add conceptual weight that distinguishes these terms.[2]

Modesty, as we mobilize the term, is fundamentally a rejection of subservience and submission.

We also distance ourselves from the idea that modesty and prudence imply some sort of repression or sacrifice. Prudence has gotten a bad rap these days! "Prudence" is commonly defined as being thoughtful about the future, a form of phronesis underpinned by wisdom and caution. We like and fully embrace this meaning!

Modest, then, is not a perfect word as we found it. Such a word, as we have noted, did not exist. Some might have instead opted for a full neologism, but we are hopeful that the connotations of modest are close enough to help us move our conversation forward. What is crucial to the argument that follows is not so much the term as the charting of the contours of an approach beyond modernity, and a name that enables sharp analytical distinctions and political mobilization.

In this context, we use the term "modest imaginary" to capture our emergent understanding of the world as it is experienced. The remainder of what follows in *Enough!* is our effort to translate this wider imaginary into an approach to political ecological practice, and what it might mean to demand and cultivate modest economies, states and livelihoods.

A modest imaginary

A modest imaginary is not a set of statements about our preferences.

Whether we (collective) like it or not, *modernity is founded on inaccurate assumptions about the world and our relationships within it.*

Wishing otherwise will not help us out of our ongoing predicaments. Discounting, neglecting or discarding modernity does not help us much either, for we must continue to understand the seductive power of modernity in order to understand why it is so hard for many to let go of, and develop a compelling alternative beyond it. Clearly naming an alternate imaginary is, we hope, part of building an ontopolitics that makes it easier to let go of modernity without feeling the loss of hope and direction that underpins much postmodern thought.

In many ways a world with less uncertainty might be preferable: uncertainty can be harmful when we lack an ability to cope with change and rely on what is expected. Uncertainty, however, is not always negative, and this is particularly true in an unequal world. Uncertainty – and its link to being illegible to others – has long been understood to constrain control, limiting hegemony over nature and others (Scott 1985). An uncertain world enables politics of resistance that modern statecraft did not expect, did not plan for, and continues to struggle to subjugate. Uncertainty, much as it challenges our modern ideas of what life ought to be, is central to creating the whimsical and serendipitous moments of our lives, and the possibility of a better future.

A modest imaginary is produced through an interplay with modest science: modest science cannot provide full knowledge, and it cannot definitively tell us how we ought to act in the world. A modest approach to science, however, does not mean we collectively ought not try to know more: we would like to know more to prepare more, and modest science helps people cope with uncertainty.

A modest approach pushes us to dwell in the ambiguity regarding limits: despite many useful calculations, the best of our science cannot tell us precisely where thresholds are and what happens if they are crossed. Conversations about limits and sufficiency are useful, but often rooted in abstract calculations and assumptions of stasis (e.g. Daly 1977; Dietz & O'Neill 2013). A modest approach means there is no final steady state of fixed consumption and population to seek; economies and ecologies will always be emergent and dynamic, responding to political, socioecological and technological change. Throughput, in the context of ecological and technological dynamism, cannot sustainably be infinite but nor is it fixed: what can sustainably be produced is a result of change, experimentation and sociopolitical decisions. In this context, we emphasize that there are constraints to what is possible, but that the idea of "limits" is far too fixed: what is possible will change with our politics, knowledge and technology. It also matters when and where change happens, how much harm has been done, as well as how much regenesis is possible.

Uncertainty undermines, but is not the inverse, of modernity: just because we cannot be certain of our knowledge does not mean that there is no utility to trying to understand our world. Drawing on designerly thinking, we might, then, replace confident plans with modest attunement to patterns and anticipation of what could be. This anticipation is not about strict statistical planning for probabilities, but instead about preparing and engaging in ways that are open, adaptable, iterative and reflexive.

A modest imaginary is prudent, but not purist: there is not a perfect strategy forward from where we are, nor do we seek a perfect world of harmony with nature and each other. Following Shotwell (2016) and Puig de la Bellacasa (2017), scientific certainty and environmental purity are not achievable goals, and indeed the pursuit of these goals has repeatedly impaired effective policy implementation. There is no primordial state of purity to return to, nor can we build a world that escapes histories of slavery, colonialism and genocide. We cannot – ought not – pursue a return to a wild past (see Cronon 1996) but instead embrace our role in tending to what Marris (2013) calls "rambunctious gardens".

There is, equally, no future in which conflict is permanently erased, no final state of harmony, no end to politics (Mouffe 2014). Instead, actions are embedded within complex webs of suffering; justice and sustainability will always be subjects of struggle through agonistic politics rather than final states to achieve. Taking this as the starting point enables ethical and reflexive action without inducing paralysis.

Entangled with, but not subsumable to, a modest science, is a modest politics. Modest politics are based on an approach to material welfare that is distinct from modernity and neoliberalism, one we tease out in the remaining chapters. Crucially, it differs from both arcadian and modernist political projects in not defining good subjectivities to be pursued through collective politics (we address this in more detail in Chapter 6).

Our modest politics centralizes "enough" as a shared socioecological goal. Enough can be defined and pursued in many ways, as the literature on sufficiency shows (Jungell-Michelsson & Heikkurinen 2022). Biophysical and social science ought to inform shared decisions, but the answers to what is enough are, above a biological baseline, fundamentally political decisions about values and trade-offs. Beyond a baseline (e.g. calories, water and shelter), enough can be understood as the outcome of a sustainably produced fair share. What is enough meat, clothing, mobility is subject to – derived from – what can be sustainably produced and fairly shared.

Modest is, then, radical and optimistic in its insistence that there can be enough for all, and that we (collective) can and ought to build a world where everyone experiences having enough.

Threads of a modest approach

In this section, we tease out three key aspects of a modest approach to help us move from an imaginary of how the world works towards more concrete politics and practices. For, let us (Mary and Tyler) momentarily return to the feeling in our introduction: we have had enough! We want change! We work to develop a modest imaginary not as ivory tower academics motivated simply to understand the world. We do want to understand it, but the *reason* we develop this imaginary is, fundamentally, to *use* it to participate in a larger collective process of envisioning and creating socioecological change.

We frame emplaced sustainability, anticipatory guiding and entangled autonomy as already-existing threads in, beyond and congruent with the wider literatures on uncertainty reviewed above. They help us to ground the socioecologies we seek, as well as the politics that might help create them, and are woven through the chapters that follow. We seek to recognize, learn from and draw on places where these threads are already present, already being spun, shaping politics and practices. We imagine these threads might be woven in unique ways across different places to produce pluriversal tapestries.

Emplaced sustainability

The idea of emplaced sustainability works across two key threads of geographical scholarship: that on human–environment relations and that on spatial politics. It suggests the ongoing importance of locale within an interconnected world, highlighting the necessity of attuning political interventions to the social and biophysical contexts of specific sites and the need to act responsibly with relation to them. We appreciate, draw on and extend the use of this phrase by Barron *et al.* (2020).

Modern approaches to sustainability that emphasize growth and development, as well as abstract science, present sustainability as a destination to be reached. It can be universally known and sought. In contrast, *emplaced* sustainability is a negotiated and always ongoing process that will be enacted differently across space (see Leach *et al.* 2007). "Place", here, is not synonymous with location but is a way of understanding where we are as nodes, sites of interlinkages within a global set of circulations of people, ideas and things (Massey 2004). Places exist across many scales, and are constituted in part through their relations to elsewhere. Moreover, "local" in the parlance of Massey that we follow here, must be understood relationally and in its connection to other places. Environmental problems exceed local scales, meaning that sustainability needs to engage multi-scalar processes. Emplaced sustainability is, thus, quite distinct from "local sus-tainability" (see Chapter 4).

Knowledge for emplaced sustainability is not simply a series of movable facts but insights generated experientially through working with, and working to change, our world (Whatmore 2009; Braun & Whatmore 2010). Thus, emplaced sustainability builds upon the knowledge of local residents, particularly those Indigenous to a place, who have accumulated lifetimes of experiential observa-tion passed across generations, regarding how local ecologies function (Berkes 1999; Kimmerer 2013). It requires rethinking the ways that we conceptualize relationships between the social and natural world (Johnson & Murton 2007; Sundberg 2014; Larsen & Johnson 2017). It also revalues the knowledge of field scientists, who relied on a depth of embodied experience and tacit knowledge grounded in the study of particular places (Golley 1993; Kohler 2002; de Bont & Lachmund 2017; Özden-Schilling, in press). It also creates new opportun-ities for reappropriating digital technologies to empower new practices of self-government and localized self-determination (Calzada 2021; Schneider 2022).

Thus, emplaced knowledge might well come from everyday embodied experiences and advanced technologies connected to the internet of things: as we work to show throughout this text, there is nothing about a modest imaginary that opposes the use of technology. While knowledge is emplaced, learning can occur across places: useful knowledge can also be generated analogically, situated in con-versation with insights from those sites with topological resonances, where people have seen both different and repeated patterns (Martin & Secor 2014). Increased mobility of people is therefore not inherently bad either for justice or sustainability. It means, instead, that people's awareness, attunement and attachments to place may be developed iteratively and comparatively across different sites.

Emplaced sustainability therefore cannot be created through abstract uni-versal scientific models nor ought it rely solely on the insights of those who live and work in a location. There is a need to both draw on longstanding traditions of thought in and beyond the academy and, in a context of socioecological change, to unlearn and learn anew from where we are. Indeed, as we live in a world of expanding climatic uncertainty, understanding patterns across different ecol-ogies will be more important than at any past point. Knowledge for emplaced sustainability can therefore be generated through dialogue extending in multiple directions – to forms of knowledge produced in different ways, across many locations and scales (Pearsall *et al.* 2022).

Practically, emplaced sustainability will likely be sought by experimenting with a diversity of technologies. Tinkering and improvisation will be crucial practices, based on a recognition that these must work with and be responsive to, rather than control, their surroundings. Practically, both unequal politics and unsustainable material flows mean emplaced sustainability is easier with shorter and more circular material flows. To speak more directly, those of us whose material lives are supported by extractions from elsewhere will need to rework the scales of our socioecologies rather than colonize elsewhere. As we elaborate on later in this text (and see Lawhon & McCreary 2020), this means enabling people in places of extraction to say "no" to unsustainable development, assuring socioeconomic alternatives and building socioecologies that can be witnessed and guided through more accountable politics.

Thinking about "place" also means recognizing that place is not a singular perspective; there are multiple coexistent, competing frames about what a place is and ought to be (Martin & Pierce 2023). A modest approach, drawing on scholarship on the politics of place, does not determine the outcome of these negotiations, but instead emphasizes just processes of negotiation. Mobility and multi-scalar politics can make it difficult to understand who ought to have a say in emplaced sustainability, and there will be ongoing concerns about whose voices ought to count beyond the scope of what we outline here (see Pearsall *et al.* 2022). Rather than resolve this concern, let us note that we are emphatically committed to multi-scalar sovereignty and that the chapters that follow are rooted in a politics that enables the prioritization of those with rooted interests in the future of a place over those with temporary, extractive relationships (see Lawhon & Chion 2012).

Anticipatory guiding

Working towards emplaced sustainability means developing multi-dimensional, non-linear plans. Here our thinking is informed by scholarly conversations on critical futurities, including feminist, anarchist and Afro-futurist work that seeks to prefigure and embody the forms of change it envisions in the world. Where modernism seeks to define and control the future, and the arcadian seeks to return to a mythic past, a modest approach involves acting in a world that is and can be only partially knowable. In this context, we propose "anticipatory guiding" as a way of shaping the world without being quite sure where we (collective) want to go, and what is socioecologically possible.[3] It means accepting that neither nature nor society can be managed and controlled without giving up on the idea of cultivating a better world. Instead, guiding is a collective process that we collectively undertake together, iteratively and reflexively working to shape our socioecologies, based on learning and adapting rather than confidently charting a path forward.

One strategy for anticipatory guiding is prefigurative politics, a longstanding anarchist approach seeking to nurture the seeds of a new world within the shell of the old, building new social and ecological relationships within movements aimed to transform dominant regimes (Dixon 2014). As Jeffrey and Dyson (2021) stress, practices of prefigurative future-making typically proceed through multiple forms of improvisation. Examples include community gardens, collective kitchens and squats, but the prefigurative also extends to establishing restorative approaches to community dispute resolution and sexual assault accountability. Prefigurative politics are imperfect, for such examples are not immune to hierarchical relations. As above, a modest approach is not purist but working towards better relations rather than refusing participation in an imperfect world.

Importantly for us (Mary and Tyler), anticipatory guiding is distinct from forms of prefigurative politics which presume we already know where we are going and what a future we seek entails (e.g. much orthodox anarchist thought has an established and universal vision of a desired final state, see Day 2006). We are inspired by feminist and Afro-futurist thought that emphasizes the need for theories and strategies of social change that embrace a world in flux (Haraway 2016; brown 2017). Drawing inspiration from the provocative science fiction of Octavia Butler (1993, 1998), feminist authors have emphasized the importance of theorizing change within a world that itself is always already changing. Instead of a conservative politics of resilience, which tries to preserve the capacity of systems to maintain their structure in the wake of crisis events, these critical strategies emphasize the importance of assessing the shifting currents of change in a dynamic world in order to better shape and direct those forces that are remaking the world (Moten 2017).

In this sense, anticipatory guiding is akin to a collective of sailors amid stormy seas guiding a boat through their relations to it, rather than a tour guide on a well-trodden path showing others *the* way. "Anticipatory guiding" is thus another imperfect term that we hope can usefully enable us to distinguish and ground our approach, iteratively working together along a shared, changing pathway towards an unknowable future.

Entangled autonomy

Our final key concept is entangled autonomy. Like emplaced sustainability, we draw on and advance recent articulations of this phrase (Clancy *et al.* 2021). Autonomy is a longstanding and emotive term, used often but, as Naylor (2017) notes, undertheorized. The idea of both free-floating individuals (woven through libertarian and anarchist thought) and isolated communities (more common in Marxian and postdevelopment thought) have been widely critiqued (Haraway 2016; Bawaka Country *et al.* 2016; Charbonnier 2021). Instead, we are relational, constantly entangled and produced through these entanglements. In this context, debates over autonomy are better understood as divergent views on *which relations* advocates seek to loosen and tighten.

The meaning of justice we adopt in *Enough!* is corollary to that found in much autonomist, anarchist and decolonial thought: justice cannot be defined by its contents nor achieved through the provision of basic rights and needs. Promoting just entanglements, then, means creating conditions for substantive choices and supporting the development of capabilities. For some, this might mean more autonomy to make certain decisions; for others, it might mean tighter weavings with others. Embracing sovereignty means it is not for us (either as authors or part of any collective) to decide.

Further, entangled autonomy includes freedom to shape the entanglements that comprise who we are. For who we are cannot be separated from the power-laden relations through which we create our worlds. Worker, investor, mother, entrepreneur, Indigenous, gay, woman, white and more are not natural categories but identities configured within political and economic milieu. They are products of historically specific modes of subjectivation. Moreover, these normative subjectivities configure particular politics and its horizons. Thus, drawing on Walcott (2009), we (Mary and Tyler) see the end of modern subjectivity as an opening, an opening to reflect on and work to create relations and subjectivities otherwise. We cannot know what kinds of personhood might arise if we (collective) were assured material security, if we no longer feared for our ecological safety. Yet the freedom to create new kinds of personhood is central to the vision we articulate through *Enough!*

Adopting this perspective has implications for economies, states and livelihoods, including the kinds of demands we (collective) make of the state. In Chapter 4, we (the authors) emphasize that entangled autonomy is not a freedom to act regardless of consequence: the always-already-existing inescapability of some degree of entanglement means collective regulation must limit negative impacts on others (ecological and otherwise). In other ways, the autonomy we (Mary and Tyler) support means a more constrained role for the state: a modest approach to justice enables diversity in a world in which the collective cannot and ought not decide for others what a home is, how to care for their children, what food they ought to eat, what schedule to follow, how much water ought to come through the taps, or how we choose to spend our time. In the chapters that follow, we further develop the idea that the state can play a role in enabling autonomy without determining the contents of what a good life entails.

Conclusion

Chapter 1 provided an overview of key debates in political ecology about the future, emphasizing the ongoing significance of modernity to contemporary debates. This chapter offers and develops the notion of a modest imaginary as an alternative view that better accords with the worlds in which we dwell. It is inspired by our reading of socioecological and social theory that is neither modern nor anti-modern, including ideas of new ecological, resilience, post-colonial, Indigenous and feminist theory. Choosing a new term enables us to fill it with content without collapsing very real differences among these literatures, and different authors within each tradition. A modest approach draws on commonalities across some versions of these literatures, mindful that there are surely further sources from which differently situated scholars might draw inspiration.

Having outlined our conceptual framing and situated it in ongoing debates over political ecological futures, in the remainder of the book we work through what this means for both the futures we want and the politics, policies and practices that might help us get there. After a brief interlude on universal basic income, in the next chapter we start by examining what modest economies might entail. We then consider how this might be enabled by a modest state (Chapter 4) and modest livelihoods (Chapter 5).

INTERLUDE: THE RADICAL POTENTIAL OF A UNIVERSAL, UNCONDITIONAL BASIC INCOME

By the time you read this book, the idea, policy and practice of cash transfers might not need much explaining. In a few short years, cash transfers have leapt from being a controversial approach to development primarily in Africa and Latin America to a practice found in cities and countries across the north and south. The Covid-19 pandemic accelerated this trend, but the expansion of public conversations and practices of cash payments was already happening.

This rise is rooted in a frustration with conventional approaches to development and welfare, and a sense that state- or NGO-led projects often did not meet the needs and aspirations of recipients. Many governments in the global south began to shift away from undertaking projects towards directly giving money to the poor. Cash transfer programmes in Brazil and South Africa are notable early examples: they are conditional in that they provide money to targeted groups based on who is deemed "worthy" of state funding, but there are no restrictions on how the funds can be spent (Hanlon *et al.* 2010). Cash transfers gained attention in the context of debates over automation in the global north (Stern 2018; Yang 2018), and became normalized through Covid-19 payments from Sri Lanka to Ghana. As just one measure, Gintilini (2021) suggests that one-third of Covid-19 social protection was "cash transfers reaching over 1.1 billion people, or 14 per cent of the world's population".

"Cash transfers" is a broad term, and thus far all state transfer systems have entailed some form of conditionality: they are offered to certain populations, living in certain places, for a certain length of time. Sometimes, eligibility is dependent upon certain actions, like children's school attendance or submitting job applications. Many cash transfers are not expected to be enough to ensure a basic livelihood without supplemental income (Ballard 2013). While momentum is growing in the present, the idea of some form of state-granted social transfer has come and gone from public discourse many times. Early examples can be found in Europe and North America in the eighteenth century, where discussions of human rights included the consideration of economic rights particularly for those unable to work (Bidadanure 2019). The idea has been advocated for across the political spectrum, making its politics somewhat difficult to categorize.

Those interested in economic justice often frame basic income not only as a mode of economic redistribution, but as an ethical claim to collective shares in our intellectual and ecological inheritance, a share of the material resources provided freely by nature (Ferguson 2015). How basic income is framed and justified can impact its support: Fouksman's (2021a: 286) powerful reflections on basic income in Namibia are telling here. After extensive justifications of work from her respondents, she concludes "those who were initially reluctant to support a basic income grant became enthusiastic when I described it as not a government 'grant' but a 'dividend' or a rightful share of the country's wealth" (see also Fouksman 2021b).

In this book, we emphasize the merits of a cash transfer that is truly *universal*, meaning that there are no conditions attached and that no one will have to apply for, nor work for, nor undertake any actions in order to obtain it. A universal programme has political and pragmatic benefits: it has no normative politics beyond redistribution and poverty alleviation, it reduces the possibility that recipients will be missed, it eases administrative burdens. It is also *basic*, meaning it is enough to live on without a supplemental income (see Calnitsky 2017; van Parijs & Vanderborght 2017; Standing 2020). What is enough is a function of many things, including social expectations, politics and the arrangements through which material goods are provided. The low-end value of what qualifies as "basic" would, for example, be lower in a place with free health care and free education, and even lower with free or subsidized basic water and energy. It also changes if our wardrobes are durable and social expectations normalize repeated wear, if we eat less meat and rework food subsidies, grow some food and eliminate the excesses of lawns. A UBI can be higher if our politics enables the collection of more taxes. Such practices and politics may shape calculations that provide a numerical value of what is enough; however, key to our argument throughout this book is that these imagined practices provide a guide for an estimated sufficient income, *not a set of norms of ideal consumption*. The fungibility of cash – discussed more in Chapters 4 and 5 – means welfare is provided without enhancing the biopolitical power of the state, enabling recipients to construct their own livelihoods (Lawhon & McCreary 2023).

The UBI we support is *redistributive*, derived from taxes on profitable industries and the rich: the result would be a roughly equal income for those in the global middle, and a clear net gain for the poor. Such a programme – like all social gains – is unlikely to happen in a single collective moment: the history of social democracy has shown that partial programmes are often the starting point on the way towards wider benefits. A global UBI is, thus, a long-term goal that can be built incrementally over time from creating new cash transfer programmes, expanding existing ones, and building on existing redistributive practices and political demands (e.g. payments for ecosystem services, compensation for north-south ecological debt, reparations for legacies of slavery and land theft, see for example McCarthy 2004; Curley & Smith 2020; Fletcher & Büscher 2020; Ramirez 2020; Klein & Fouksman 2022; notably, a UBI does not necessarily preclude other targeted forms of redistribution). Once established, it is difficult for democratic states to take away social benefits, and gains often generate political momentum towards expansion. Demands to expand conditional cash transfers, and pressures to retain pandemic payments, suggest an increasing normalization of, and desire to extend, cash transfers. We, along with many others, believe incremental expansion is the most likely pathway towards a global, redistributive UBI (Webb 2021; Torkelson 2022).

Cash transfers have been widely correlated with positive social and political impacts that extend beyond the economic domain (Bastagli *et al.* 2016). Previous and ongoing experiments with cash transfers suggest that the adoption of a UBI could improve health and increase educational attainment, as well as reduce stress, crime and violence (Forget 2011; Ruckert et al. 2018; Arnold 2018; Calnitsky & Gonalons-Pons 2021). Moreover, while increasing wages can also result in positive benefits, the effects of increasing income from cash transfers do not mimic those of higher incomes from labour. For example, increased incomes from wages correlate with increased consumption of what are called "temptation goods" (e.g. alcohol and tobacco), yet cash transfers actually *reduce* such consumption (Evans & Popova 2017). Further, and more significant for our wider argument, Owusu-Addo *et al.*'s (2018: 689) review of the literature found that cash transfers "have improved social cohesion in diverse ways through improved security, increased sense of belonging, enhanced social status and state-citizen social contact". Our point here is not to critique particular forms of consumption but instead to insist that the wider socioeconomic and political implications of cash transfers and increased wages are not the same.

Many interested in radical economic change have framed a UBI as an ameliorative policy, one that reduces the ills of capitalism and pacifies citizens. A central argument throughout this book is that a redistributive UBI is unlikely to produce political stasis, and instead might well catalyse and underwrite a just and sustainable future (Haagh 2019; Lawhon & McCreary 2023). First, in an argument we make fully elsewhere (Lawhon & McCreary 2020), eliminating the reliance on labour for income makes it easier to say "no" to extractive developments in particular places, subtending the ongoing and very real conflicts between environment and development. We return to this point in Chapter 5 where we consider the relative merits of basic income and a jobs guarantee.

Second, a UBI is also already a politically acceptable way through which to move billions of dollars annually away from capitalist economies towards non-capitalist ones, a point we elaborate on in Chapters 3 and 5. As Wright (2004) has argued, a UBI can serve to bolster the bargaining position of workers by creating a "permanent strike fund" that enables them the choice to engage in paid labour. However, the anticipated effects of a UBI extend beyond formal capitalist relations. It increases the possibility for engaging in decommodified activity and expanding economic practices beyond capitalism (Wright 2004). The economic security provided by a UBI can enable experimentation with alternative economies. While there are no explicit conditions, it is worth noting that the forms of economic participation fostered by cash transfers have tended towards greater support for small businesses and, for some, a reduction of paid labour (Lawhon & McCreary 2023; see Chapter 5 for a defence of this reduction). Further, economic security combined with more free time enables greater participation in the time-consuming tasks of creating democracy and emplaced sustainability.

Many of the authors cited throughout *Enough!* have noted their support for a UBI, at times in an offhand fashion. For example, Erik Olin Wright is a widely noted socialist advocate (2004), Naomi Klein notes her support for basic income in *This Changes Everything* (2014; see Box 3), Jason Hickel also suggests compatibility between degrowth and basic incomes (2016). These authors largely present a UBI as part of a basket of positive policies that can be demanded from the state. We are sympathetic, yet believe there is much unexplored in such a framing. Throughout *Enough!*, we work to tease out this underemphasized potential, considering a UBI as an example of a distinct mode of statecraft (Chapter 4) and a potential catalyst through which to underwrite the changes we seek (Chapter 5).

In the world we have, a UBI is our best chance for politically acceptable economic redistribution that can enable radical change.

To be clear, a UBI does not guarantee that money will be spent building and supporting other forms of economic organization, or that people will use their time to engage in democratic practice. Ours is not a deterministic argument about what would happen. But – and here is where our optimism comes through – we do believe that people want change. We believe that, if assured greater security and freedom, with a vision supported by social movements and public conversation, most people would choose to spend their time and money in ways that build more just and sustainable economies.

This proposition is a gamble, a gamble based on the belief that we can build a vision that is compelling enough for people to choose it. It is a risk. But in this way, it is no different than the radical visions that precede and parallel ours: radical change cannot be justly forced on people! Instead, we (collective) must build a shared story that people will freely choose (Lawhon & McCreary 2023). It is towards this end – to outline the kinds of economies, states and livelihoods that might be created through a modest approach and underwritten by a UBI – that we write *Enough!*

\

3
A MODEST ECONOMY

One of the main stories explored in *This Changes Everything* (Klein 2014; Lewis 2015; see Box 3) is that of Mike and Alexis, a goat ranching couple in Montana. The film follows the landowners through their oil-soaked fields as they witness the impact of pipeline ruptures. We hear Mike and Alexis's heartfelt lament, and their concerns clearly go beyond regret for an economic loss. These ranchers are presented as sympathetic characters, seemingly part of the sustainable society that Klein and Lewis call us to support.

Analytically, however, there is something unsettled here, something ambiguous that came up time and again when Mary taught with this film. Capitalism – clearly presented in the film and book as part of the problem – is generally understood to be a system of private property. And Mike and Alexis are, presumably, owners of private property. Conventional critiques of capitalism root their concerns not just in the particularities of certain industries – oil/bad, ranching/good – but in the *mode of production*. Orthodox Marxism places small landowners and artisanal producers like Mike and Alexis within the "petite bourgeoisie", allies of the system with interests counter to that of the proletariat.

What are we do to with this ambiguity? What kind of political economic analysis enables us to separate fossil fuel corporations from petite bourgeoisie like Mike and Alexis?

We are far from the first to consider the complicated role of small ranches, farms and family businesses in more just economies (see Scott 2012; Levien *et al.* 2018). Particularly in the global south, there has been extensive debate over what an agrarian socialism might entail and the utility of Marxian thinking in advancing it. Considerations of the radicalism – or not – of the Brazilian Landless People's Movement (MST) are exemplary here: the unifying demand of MST is a redistribution of land, a taking over of the means of production (Wolford 2010; Vergara-Camus 2014; Carter 2015). The demands of contemporary landless peoples are often made with an awareness of disastrous histories of collectivization by the state, campaigns that often forced changes on farming communities. The outcome of land redistribution is primarily not a move from private to common property, but a shift from large private landholdings to small ones, accompanied by a shift from export-oriented production to subsistence and local market production. Similarly, socialist thinkers in the global south have been forced to reckon with the significance of the informal sector, broad economic domains that remain beyond the reach of state licensing and regulation. It is a sector filled with variations on private enterprise. Efforts to suppress informal businesses have been widely seen as regressive and anti-poor, and efforts to turn them into collective enterprises met with variable outcomes (Lindell 2010; Mohan *et al.* 2018; Tucker & Anantharaman 2020).

What does all this mean for how a just economy can be identified, named, defended and demanded? Such questions trouble orthodox political economic labels and analysis, blurring established narratives of allies and enemies as well as appropriate courses of action. At stake here is the form of a just economy, including the question of whether a just economy can include privately owned enterprises, and if so, what might distinguish this economy from a capitalist system driven by profits and growth.

In what follows, we first provide a brief explanation of canonical debates in political economy over capitalism and socialism, then point to their shared rooting in modernity. We then draw on postcapitalist and postdevelopment scholarship that helps us, at times, to think about non-modern economies. Yet, as is true throughout this text, we insist on the need for an affirmative word, one not oriented towards not-being-capitalist and not-being-developmentalist. We also point to some limitations of postcapitalist and postdevelopment thinking, including its arcadian tendencies and limited explanation of how to induce change. Then, drawing on the ideas of entangled autonomy, emplaced sustainability and anticipatory guiding, we consider what a modest economy might entail and how it might be built from the world we have. This modest approach centres on ensuring a balance of power and sustainable material sufficiency, resonating with key concerns in many non-modern political economies, such as gifting societies. We conclude this chapter by showing how cash transfers, and particularly a UBI, might enable and underwrite a modest economy. Crucially, we argue that a UBI reduces the risks associated with, and enables the pursuit of, a range of non-capitalist enterprises without increasing the biopolitical power of the state (see Lawhon & McCreary 2023). In sum, the modest approach we develop here is not a search for an ultimate and final economic state, nor can we clearly and universally define the contours of all just and sustainable economies. It is about shifting our gaze and questions, as well as our practices and demands.

Capitalism, socialism and modernity

Broadly speaking, debates over political economy, justice and sustainability have long centred around the appropriate form and role of property and profits. Capitalism and socialism feature prominently as important, highly emotive lodestars to orient discussions of economic systems. There is and has been much value to such discussions. They help us to grapple with the relative merits of different property regimes, understand our values and name our allies.

And yet, it is easy to see the limits of experiences with both systems: the freedoms promised by capitalism have little value for most people in the context of widespread inequality and ecological crises. The planned economies of socialist states led to grave injustices carried out in the name of a better future and stifled the diversity of human aspirations. Both contributed to widespread environmental harm. Further, no economic system has ever existed in a pure form: history is not a teleological set of stages of replacement. Real-world economies are all mixed, comprised of myriad types of ownership, with coexisting and competing logics (Gibson-Graham 1996, 2006). And, as we detail more below, modernist versions of capitalism and socialism are poorly equipped to grapple with the uncertainties and ambiguities outlined in Chapter 2.

Capitalism versus socialism may be the most familiar entry point for political economic debate. It is, however, not the only way through which to understand economic justice.

In this chapter, in keeping with the approach developed throughout this book and drawing on the work of many heterodox political economists, we point towards a common set of assumptions shared by orthodox versions of capitalism and socialism: both are underpinned by modern notions of rationalism, empiricism and scientific management (Lushaba 2009). This is not to collapse distinctions between them, nor to suggest that all capitalist and socialist theories deploy modernity in the same way. It is crucial for the argument we develop in this book, however, to emphasize the central, foundational role of modernity in both capitalism and socialism.

We write this chapter mindful that there is an increasingly capacious set of ideas proposed under the name of "socialism", underpinned by visions of the future that look very little like modernist state-planned socialism. This non-modern way of doing socialism has long roots, but is increasingly prominent in contemporary conversations. Its difference from orthodox modernist Marxist socialism is not always explicit and in the text that follows we consider aspects that are ultimately compatible with the modest approach we develop here. As with our analysis of degrowth, the key analytical distinction we work to make here is between modern and modest approaches.

The origins of modern economies

The political economic logic of capitalism is rooted in, and deeply entwined with, modernity. Capitalism is founded on the modernist idea of rational individuals whose economic decision-making relies upon access to information and transparency of knowledge, as well as idealized models of market stability established through conditions of economic equilibrium and perfect competition (Fleming 2017). This is the economics taught in most introductions to the subject, presented as "truth" rather than one approach. I (Mary) remember sitting through lectures at the University of Kansas twenty-some years ago, watching the charts and graphs fly by on slide presentations. I loved the clean logic of this version of economics, and the clear and singular solutions to the exam questions. Here were answers to what should be produced and who should be able to access it, provided by mathematics! We confess to wishing, sometimes, the answers were this easy, logical, universal and justifiable.

But if you've read this far, you know we don't think the answers are this easy.

How did some of us (collective) come to believe that this is the way the world works, and ought to work? And are we, finally, in a moment when this way of thinking may no longer be tenable, may no longer be able to be defended, should no longer shape how we think about what economies are and ought to be?

We (Mary and Tyler) skip a lot of history here, but let us zip through some highlights. When it emerged, modern capitalist thought was a powerful contrast to the dominant economic ideas and practices of Europe (and elsewhere). Feudalism was underpinned by moral narratives and hierarchical social relations. It was a profoundly unequal system, but one in which there were mutual obligations between serfs and lords, husbands and wives, parents and children. In anti-feudalist movements, people variously sought to transform or break their bonds to social structures of obligation. These were revolutionary ideas, compelling to many who were constrained by the feudal regime and understood to have liberatory potential.

This history is most well-documented in Europe, but there are parallels elsewhere as well: for example, many African people saw participation in modern economies as a means to overcome kyriarchal political economic relations. Young men and women anticipated that wage labour would free them from reliance on chiefs and customs, enabling greater autonomy (Cooper 1996).

The shift away from hierarchical, moral economies like European feudalism was not linear, and what form the subsequent political economy would take was long unclear and deeply contested. Economic historians have increasingly pushed us to reconsider minor histories too often forgotten. In Europe, among those revolting against feudalism were peasants making demands for a redistribution of power and wealth (Poulsen 1984). These rebels claimed ownership of the lands where they lived and worked and demanded shared access to common lands. Some of these people did, sometimes, work for an income that would supplement what was grown and gathered. But rather than accumulation, they largely worked until they had enough. And then stopped working.

What might have happened if this economy had been allowed to persist, we cannot know.

The pushback against it – not only by the feudal lords, but also by emergent capitalists – was fierce and unfolded over multiple generations (Fairlie 2009). Ultimately, these rebellions and experiments were largely quelled, typically through violence and forced enclosure.

This is also the history of capitalism: it emerged not simply and solely out of feudalism, but through the suppression of demands for a more egalitarian mixed economy in Europe through the Renaissance period.

Interestingly for us (Mary and Tyler), we (collective) have no label for the kind of economy that many of these anti-feudalists sought. It was underpinned by an egalitarian vision of distribution and access, but was not socialist in any modern sense of the term. In recent years, political economic scholarship has sought to more carefully attend to this history and imagine how it might shape ongoing visions of political economic change.

Let us, for now, return to the history. Alongside anti-feudal demands for a more egalitarian mixed economy, there was a countervailing line of capitalist development. Through the enclosure of common lands, landlords sought to maximize their opportunities for private accumulation, and thus "capitalism was advanced by the assertion of the landlords' powers against the peasants' claims to customary right" (Wood 1999: 118). Private property was created through violence and enclosure. Peasants were dispossessed of their means of subsistence. They became legally free and independent of feudal bonds, but also propertyless, relegated to selling their labour in return for wages.

Privatizing large landed estates created a class of people who needed to work for a living. This work was no longer part of the feudal moral economy, nor was it supplementary labour undertaken by those with longstanding historical tenure rights. Instead, for the first time in human history, there was now a class of displaced people who needed to make a livelihood by directly selling their labour to others, then use that money to purchase commodities to sustain themselves (Harman 2004).

And this happened at just the time when the Industrial Revolution was taking off, a revolution funded by the profits of colonial exploits and bolstered by resources from these colonies. An economic revolution that needed workers.

In this new political economy, prior social relations and obligations were displaced by a modernist rationality underpinning a new political economy. In practice, this was, crucially, never a full or complete displacement: generations of peasants, Indigenous peoples and producers across the north and south have always maintained other economies and relations alongside, outside and in resistance to capitalism. As we detail in Chapter 5, the subjectivity of the modern worker had to be actively created, and often failed to materialize. Even for those immersed in capitalism, other logics continue to coexist (Gibson-Graham 1996, 2006). However, the new political economy induced monumental change both materially and cognitively, as modern thought mythologized the norms of capitalist relations.

Subsequent centuries of prevailing economic thought, spread through colonial flows of knowledge, have been rooted in the idea of *Homo economicus*. In the context of growing urbanization and industrialization, underpinned by colonial exploitation, emergent forms of scientific management sought to rationalize labour processes to maximize productivity in the centres of empire, subjecting workers to systematic regimes of industrial engineering. These processes had some parallels globally, although colonial labour relations were more often underpinned by violent exploitation and coercion that relied on the invention and social engineering of racial categories (Cooper 1996).

Socialism came into being as a way of working against the inequalities and alienation of European industrialization. Marxists sought to collectivize productive forces, expanding the project of modern management to direct the productive capacities of society towards ever more economic efficiency. They sought economic power through a politics of the ownership of the means of production, not distribution (Ferguson 2015; see Burawoy 1985). To state directly: Marx and his immediate followers sought to collectivize modern industry without fundamentally reworking the modernity on which it was built. They aspired to create socialist states that would collectively prescribe development, deciding what gets produced and where.

In Western Europe, socialists sought to expand state authority to plan economic relations, including introducing programmes of incentives and regulations to direct economic patterns. Elsewhere, twentieth-century state socialism emerged in a more expansive way. This included ownership and control of production by centralized state authorities that aimed to define public needs and then uniformly meet them. As it was envisioned, the socialist state would induce production of items deemed important, and reduce or eliminate production (as well as importation) of items deemed unnecessary. To control the price of key commodities, socialist states instituted marketing and purchasing boards, seeking to concentrate economic leverage to, respectively, maximize or minimize prices at different points in the value chain. The ongoing impacts of such practices can be seen, for example, in efforts to control the price of staples such as wheat in Egypt (Barnes 2022).

These historical and contemporary efforts are premised on a series of modern assumptions, most centrally in the capacity of the state to know and accurately predict production capacities and social needs, as well as the ecological costs associated with different development trajectories. In short, most socialist critique of the twentieth century was directed at capitalist patterns of ownership and profits, not modern production itself. Alternatives sought to change the political economy of production, not challenge modernity.

This is not the end of the story of socialism, however, and we pick up this thread again below.

The limits of modern economies

The modern approach to the economy has led to massive increases in productivity over the past few hundred years. These increases have improved the material standard of living for many globally and particularly in advanced capitalist countries. Our critique of modernity does not discount these very real improvements.

Modern economies have, however, been subject to extensive critique.

The industrialization of Europe was deeply entwined with colonial exploitation and oppression. This point continues to be politically controversial, yet empirically difficult to contest. It is, ultimately, unknowable how European economies would have unfolded without the contributions of materials and labour from the colonized world. It is equally unknowable how economies across the globe, and particularly in the global south, would have unfolded were they not colonized. It is clear that global inequality, and particularly inequality between the north and south, was marginal at the start of the sixteenth century (Wallerstein 1974; Thompson 2012). While material production and consumption have risen across the south over subsequent centuries, the pace of real material improvement has paled in comparison to the significant rises elsewhere, producing a profoundly unequal global economy.

Within the story of modernity, enduring colonial discourses rationalize such inequalities (see Boxes 1 and 2). Colonialism was understood to be a "civilizing mission", bringing modernity to the global peripheries. Development became a narrative through which to organize countries (or "civilizations") into different stages. Following formal decolonization, discourses of development continued to depict the global south as in need of aid to reach a state of modernity (Escobar 1994). This way of thinking was, in many ways, shared across capitalist and socialist approaches to development.

There were and are, of course, abundant challenges to colonialism and its legacies. One line of critique appropriates European discourses of modernity, pointing to their exclusions and calling for a more inclusive imagination of liberty (Gilroy 1995). Colonized populations contest the colonial ordering of

benefits, and liberation movements often demanded expanded political considerations of who counts and holds the rights associated with modern citizenships.
This also included demands for more equal trade, labour relations and access to
the outputs of modern economies (Lushaba 2009). Following formal decolonization, there continues to be critiques of the exclusions associated with neocolonial economic relations.

Contestation did not only happen *within* the storyline of modernity, however. Others opposed modernity more directly, including criticizing its role in
shaping political economic thought and practice (Barchiesi 2011; Joseph-Gabriel
2019; see Chapter 1). European capitalist and state socialist regimes were also
subject to internal critiques of the alienating and ecologically destructive effects
of modern economies. Artistic critiques of modernity in the early twentieth
century, such as Surrealism, challenged the rationalist excesses of the capitalist
economy and how it suppressed the creative spirit. Within the United States,
early autonomous Marxists, such as C. L. R. James, Raya Dunayevskaya and
Grace Lee Boggs challenged the dehumanization of the factory line, critiquing
alienation at the point of production under modern industry in existing capitalist and communist regimes (James *et al.* 2013). James also extended these
critiques to broader challenges to the group-differentiated forms of exploitation
and deprivation characteristic of racial capitalism (Robinson 1983). In advanced
capitalist countries, these critiques eventually consolidated into the rebellions of
1968 (Lefebvre 1969; Kurlansky 2005).

In the planned socialist economies of the global north, resentment of state
prescriptions and the bureaucratic management of everyday life seeded uprisings
that eventually overthrew the Soviet Bloc. Across the north and south, socialist
states had too often assumed that the needs of an idealized universal worker
could be used to approximate those of the broader population. Moreover, with
imperfect knowledge, centralized authorities and corporate planning created
widespread social crises. The hubris of scientific management and planned economies was mistaken – people are less uniform, rational and docile than assumed,
while social and environmental processes are more complex than authorities
could accurately model (see a more detailed review and citations in Chapter 4).

Finally, environmentalists, too, have provided fundamental critiques of
modern economics. While some have advocated for modernist solutions
through better integrating ecosystem services into economic calculus, critical
scholarship has repeatedly demonstrated that ecological relations cannot be
effectively managed as commodified resources (Robertson 2006, 2012; Ernstson
& Sörlin 2013; Sayre 2017). Political ecologists have shown, in case after case,
that capitalist economies have only ever partially addressed social and environmental concerns, and regularly exacerbate the uneven distribution of socioecological costs and benefits associated with development. State socialist economies

too have attempted to extract from, calculate and rationalize nature in ways that have created widespread ecological harm (Shao 1986; Pryde 1991; Gille 2007). In short, the modern approach to nature is a poor representation of what we can know about, and the extent to which we can control, nature.

Such concerns with modernity have produced two divergent responses. The first is neoliberalism, a way of thinking built on a partial rejection of modernity (Chandler 2018). Advocates of unregulated capitalism, such as Hayek and Friedman, emphasize the fundamental limits of knowledge, yet retained the idea of a rational individual. While the modernist political economy sought to produce a universal subject with generic consumption habits, neoliberalism celebrated the expression of diverse consumer preferences through variegated markets. We elaborate on neoliberal statecraft and efforts to solve environmental problems through markets in Chapter 4. The second set of responses are harder to classify in a single term, and as we noted in Chapter 2 is largely defined by what it is not (postcapitalist and postdevelopment). We review this in more detail in the section that follows.

Arcadian versions of postcapitalist and postdevelopment economies

Advocates of economic diversity have responded to the failures of capitalism and socialism in a variety of ways, emphasizing the importance of practices that enact economic difference and re-embed economic relations within the context of community needs. At times, there is an arcadian thread throughout these that romanticizes community relations, small-scale operations, artisanal production and human labour. In this section, we emphasize this thread to demonstrate its resonance with the arcadian environmentalism outlined in Chapter 1. Yet like with socialism and degrowth, there are many versions of postdevelopment and diverse economies, including scholarship that has sought to distance itself from these arcadian tones. We pick up on these differences in the sections that follow.

Postdevelopment thinking is explicit in its rejection of modernity, and particularly recommendations for the conjoined approach of modernization, globalization and industrialization (Esteva & Escobar 2017). Others, primarily in the global north, writing from what has variously been called postcapitalist, community and diverse economy approaches, similarly have sought to foster economies that are diverse and rooted in community relations. As Gibson-Graham observe (1996, 2006), despite the prevalence of capitalism, other logics have always coexisted: alternative moral economies and social relations were never fully displaced by capitalism. Similarly, despite hundreds of years of capitalist relations in the global north and south, state and common ownership were never fully displaced. There is increasing dialogue and, for many, a sense of convergence across these literatures and a shared motivation to attend more carefully to existing alternatives (Gibson-Graham 2007; Escobar 2015).

Postcapitalism and postdevelopment draw explicitly on poststructural thought which emphasizes difference and the complexities of knowledge. Each of these literatures has worked to imagine ways of fleeing or building in the cracks of capitalism, prefiguring or creating "lines of flight" towards other worlds. Iconic examples of the kinds of economic activities studied and supported by advocates of diverse economies include community gardens, farmers markets and cooperative grocery stores, but there is a growing push to think beyond small-scale, local activities towards wider processes of embedding economic activities writ large (Gibson-Graham *et al.* 2018).

The ideas and examples written about here provide a necessary counterpoint to the modernist presumptions of mid-twentieth-century corporations and planned socialist economies, highlighting existing and potential contributions of collectives and commons as alternatives to the state and the self-maximizing private capitalist. They also help us to see the limits of neoliberal theories that rely on a modernist conception of individuals acting freely through processes of economic exchange. For Hayek and his followers only partially rejected modernity: they recognize that imperfect knowledge within a dynamic world shapes economic relations but retain the modern presumption that the individual forms the universal foundation for economic activity (Chandler 2018). Moreover, they assume that the rationality that governs choices within the economy is self-maximization: that individuals are possessed by a modernist desire to make decisions for their own economic advantage and possess the necessary self, social and worldly knowledge to do so. But as those writing about postdevelopment and diverse economies have repeatedly shown, not all valued forms of interaction can be translated into commercial exchanges.

We see much merit in literatures emphasizing economic diversity, logics outside of capitalism, and the valorization of economic experimentation, broadly speaking.

There are, however, limits to the arcadian line of thought that is, at times, embedded in diverse economic and postdevelopment scholarship. Analytically, it remains difficult to understand precisely what counts here as part of the story of "yes". We introduced this chapter with Mike and Alexis, private property owners who, in orthodox socialism, would be considered part of the problem. A narrative of diverse economies enables us to place them more squarely within the worlds we want, yet broad terms like "mixed", "heterogeneous" and "diverse" do not enable differentiation between more and less just mixtures.

Further, visions of sustainability and justice premised on arcadian ideas of community harmony belie existing socioecological entanglements and hegemonic power relations. First, participation in diverse economies of high-quality artisanal products is difficult for those with limited economic resources (Aguilar 2005). Creating such economies as producers is risky: there are often high upfront costs and much uncertainty associated with such investments. Participation as consumers is equally difficult as goods produced through more just and sustainable relations are typically more expensive; "ethical" goods thus appeal to consumers with a clear knowledge of the production process, sufficient disposable income and a willingness to spend their money accordingly. Often purportedly "ethical" consumers' greater wealth and levels of consumption have entailed a larger rather than smaller ecological footprint (Checker 2020). On both counts, advocates of diverse economies have been critiqued for their elitism.

Conversely, celebratory accounts of informal economies have been challenged on the basis of obfuscating underpinning structural inequities. Often more informal community economies rely heavily on highly exploited and even unpaid labour that is racialized and gendered (Samers 2005; Fickey & Hanrahan 2014). Moreover, those marginalized people that turn to alternative and informal economies sometimes do so not because of community harmony, but precisely because they have been excluded and abandoned by the mainstream economy (Bledsoe *et al.* 2022). Without some kind of economic redistribution, it remains difficult to see how community economies can address and override existing inequalities.

Further, thinking about community economies provides little insight into how existing inequalities might be redressed. Lines of flight and experiments are meant to prefigure alternative worlds, but there is no ultimate harmony to seek and no "away" to which we can escape. Individual escape may, for those who get value and joy from entanglements, be utterly undesirable; communal embedding may be equally problematic for those who reject or do not conform to social norms. Postdevelopment literatures too have emphasized possibilities for more just ways of being in the world, yet also have been critiqued for reinforcing nativist ideas of belonging and insufficiently responding to material inequalities.

Thus, subsuming economic relations to community or social moral codes is not sufficient criteria through which to differentiate just and unjust economies: as economic historians have shown, feudalism was a deeply moral economy that subsumed economic relations to social norms. It was not just. For all its many (many!) flaws, capitalism enabled some people to escape from some unjust moral obligations. Islands that offer some shelter from particular pressures may exist, but these may well rely on or exacerbate other forms of inequality. Non-capitalist enterprises often rely on non-waged labour (including family and friends) both in a postcapitalist spirit and to subsidize what are generally more labour-intensive processes. This creates complex forms of sociability and community, but can equally be understood as extractive and exploitative. Critics of capitalism have pointed to the tendency of capitalism to draw in labour that is undervalued or made invisible in order to increase profits, and there is very real danger in mirroring these processes under the name of more just and diverse economies, even if the reasons for economic collaboration and participation are blurrier.

We do not wish to imply here that community economies are somehow feudalist: they are not. But it is crucial not to romanticize or universalize the benefits of subsuming economic relations to social ones in a world that continues to be underpinned by kyriarchal social norms. An explanation of just economic relations must entail more than a prescription for the subsumption of economic relations to social ones.

More broadly, we are also concerned with the theory of change that underpins postdevelopment and diverse economies literatures. Even if islands of more just socioecological economies were possible, there are real reasons to doubt whether such economies can be adequately accelerated in the world we have without more substantive support. As many have argued, and the growing influence of capitalism over recent decades seems to evidence: in the world we have, it is difficult to imagine diverse economies successfully outcompeting and displacing capitalism. Relocalizing and creating just conditions of production, as well as making products sustainable, is quite simply more time-consuming and expensive.

Concerns with arcadian postdevelopment and postcapitalist economic approaches take on even greater urgency in the context of the ongoing wave of technological change, automation and technological disemployment (see Pierce *et al.* 2019). Much of the diverse economies literature valorizes particular forms of artisanal labour. It is likely that ongoing technological change will only increase the cost differential between goods produced with more and less labour. The risk here is twofold: the first is missed opportunities, creating a world in which people spend their time working for an income and struggling to make time for emplaced sustainability and participating in democracy. Building emplaced sustainability and democratic participation – and ensuring time for them – are crucial to the wider argument we have developed here. More substantively, we find it unlikely that a world that relies heavily on local, artisanal labour would be compelling. It is, as many have argued, difficult to convince many to believe in and strive to create such worlds.

In sum, in the world we have, modernist and arcadian economies are unlikely to produce the good green world we seek.

Box 4 A strategy for getting there: eroding capitalism, building alternatives

Within the classic revolutionary Marxist tradition, the state under capitalism is an apparatus serving dominant class interests, establishing the legal basis for regimes of exploitation and dispossession. In this version of radicalism, structural inequity is understood to be rooted in capitalism, which sets the horizons for political debate in ways that foreshorten possibilities for fundamental change. Engaging the existing capitalist political structure only reinforces its legitimacy and hold on society. In order to transcend its limits, revolutionary change is required, overthrowing the established capitalist regime and replacing it with a socialist state. Marx suggested that the revolutionary capture of the state was a necessary step in the overthrow of capitalism, and his grand view of history held that the state would ultimately wither in an eventual shift to communism.

While Marx's own postrevolutionary prescriptions were vague, his writings have been consistently used to justify the need for a strong, centralized socialist state until a mythic time when capitalism has been abolished globally. In practice, this rationalized the existence of socialist regimes with a deeply authoritarian character through much of the twentieth century. The late twentieth-century dissolution of the Soviet Union largely ruptured established dogmas about the revolutionary capture of the state as the means to transition to a final just society (although, of course, the rather large Chinese exception remains).

The fall of the Soviet Bloc provided increased heft to those who sought to approach the state not as a singular body to capture but as contested terrain where movements could strategically advance legislative reforms and policy shifts. We build on this position to develop a modest approach to the state in Chapter 4.

In this context, the kind of change being sought by those interested in justice and sustainability is increasingly understood to be progressive and incremental. Not all change, however, is necessarily radical: *amelioration* reduces the impacts of an existing system while *transformation* is what creates system change (Gorz 1967; Albert 2017; Wright 2019). Transformative changes may not be teleological with a defined endpoint in mind, but are underpinned by a narrative and strategy that starts with the world we have and envisions systemic interventions towards creating a world that is better.

If the state is a strategic site of struggle, it is not the only one: many also work beyond the state to create social change. The relationship between advocates of these two positions – working through and working beyond the state – has, at times, been fraught. But it need not be so. Recent scholarship has framed social and political activism as complementary, with both being necessary to create radical change. Wright (2019), for example, has explained the goal of anti-capitalism as "erosion" rather than "revolution". In a narrative that more fully explains the relationship between the "yes" and "no" of *This Changes Everything* (Klein 2014; Lewis 2015), Wright argues that we ought to reduce the influence of capital while growing economic presence and political influence beyond it. He frames working to regulate capital, increasing state involvement and ownership in the economy, and building non-capitalist economic relations outside of capitalism as mutually complementary tactics.

What kinds of politics might enable more support for the "yes", reducing its exclusivity and encouraging widespread participation? What might better enable not only radical activists, but ordinary people to say "no" to the kinds of extractive industries that underpin many local economies? Creating a just future means not only careful consideration of what we want, but of what activities might have leverage, might be surreptitiously able to shift resources away from those with power, might be plausible in the world we have. This is not to treat politics or public opinion as stable: changing social norms and expectations is surely necessary! But making radical demands in a world that is as yet unwilling to support them is not enough. We need to find ideas and strategies that can have traction in the world we have – working strategically from multiple angles to erode the hegemonic relations of the existing world – in order to continue making, and making space for, the just and sustainable worlds we want (Albert 2017; Wright 2019).

Enough! largely adopts the theory of change developed by Wright (2019) described above and is meant to complement and add detail to his broad vision. We work to provide a distinct language and analytical lens through which to more clearly identify the kinds of practices that comprise and enable a more just and sustainable future, including demands that might be politically plausible in the world we have.

We (Mary and Tyler) are clear that time matters: we (collective) have had enough, and ecological crises mean we must act, now! A benefit of a revolution is its swiftness, a benefit of believing in a big modern state is that action can be swift and centralized, once control of the state is won. Yet swift action towards a deeply flawed solution is an imprudent strategy! Certainly, compromises are necessary and a modest approach is not a purist one. But we ought not hold onto the promises of modernity simply because they can create change quickly if that change carries us forward towards an unsustainable and unjust future.

Towards a modest economy

Understanding the economy through the lens of modernity points us towards clearer articulation of the limits of both planned socialism and capitalism, as well as arcadian local, community-based, artisanal approaches. A modest economy is socially and ecologically embedded, but this alone does not distinguish it as more just; not all embedded economies are just! It does not reject the state nor seek to disempower it, nor does it advocate for requiring collectivization of people and subjugation to collective norms. Instead, a modest economy is subordinate to a modest understanding of justice and works to erode the hierarchies and inequalities of the world we have. Here, we consider what a just and sustainable version of a diverse economy might entail, and how it might enable the cultivation of more just and sustainable futures.

Working across the language of distributionist economy, diverse economies and anti-kyriarchal thinking, we suggest that a just economy is not defined by its mode of production but by the distribution of resources and economic capacities. This distribution is not a singular state to be achieved, but an agonistic process that will require ongoing political attentiveness to ensure the spreading of wealth and the democratic control over parts of the economy that may need to be centralized. Distributionist thought has many roots, including a social movement called Distributionism active primarily in the UK in the early twentieth century.[1] Mechanisms for distribution exist in many cultures and take all kinds of social forms, from the sharing of meat from a hunt to spreading the care of, and benefits from, cattle (Polanyi 1971; Ferguson 2015, 2021). In short, despite the focus of much political economic work on production (and increasingly, consumption), redistribution has long been widely documented as a core economic activity of societies, whether through rights-claims, sharing, gifts, taxes or charity.

In what follows, we work through three interlinked considerations at the forefront of a modest economy: technology and labour; ownership; and distribution. As is true throughout our book, this is not meant to be an exhaustive overview but instead, we hope, a starting point for future conversations.

Technology and labour

The relationships between sustainability, technology and labour have been the subject of extensive economic inquiry and analysis. From ecological economics to sustainability transition studies, there is wide and ongoing debate within which we briefly work to position our argument (see Pierce *et al.* 2019 and Lawhon & McCreary 2020 for a longer version of our position). First, we cite our way through one of our foundational starting points: the idea that there is enough for all without requiring the full-time labour of everyone (we return to the intersection of labour and livelihoods in Chapter 5).

Those concerned with sustainability have repeatedly been forced to grapple with the implications of curtailing some economic activities for employment (e.g. Bhalla 1992; Tienhaara & Robinson 2022). While on the whole the broad movements towards more sustainability have not negatively impacted employment, it has had serious, undeniable impacts in particular places for particular sectors and people. It is not only capitalists with much to lose in a transition towards sustainability: many local economies, including workers and secondary industries, rely on unsustainable, extractive industries. As we have written in more detail elsewhere, the geographies of these economic struggles have created nearly intractable conflicts between environmentalism and employment (Lawhon & McCreary 2020).

Such concerns are compounded by a parallel trend that is rarely simultaneously considered: automation. Technology has long been used as a substitute for human labour, and for most of human history new jobs have emerged in the wake of technological change. Yet technological displacement has resulted in significant social and geographical change (Bix 2000).

In the context of both sustainability transitions and technological displacement, so far most studies suggest that there are still plenty of jobs, but not all places and people end up better off.

While history is a useful indicator, this book is based on the idea that there needs to be radical change: we ought not assume that the way things did happen is how they will or ought to continue. Many, ourselves included, believe we have already entered a truly unprecedented technological era. It is unprecedented in two key ways: technology is capable of displacing even highly advanced human labour (Pierce *et al.* 2019), and the labour of everyone is no longer required to produce enough for all (see Ferguson & Li 2018)

Adding these two insights together presents profoundly unique circumstances: *at roughly the same era as humanity became able to provide enough for all, it stopped needing everyone's labour to produce this.*

In the capitalist world we have, technological change has already produced massive unemployment. South Africa is a particularly trenchant example here: hundreds of years of its early colonial history were shaped by policies that sought to *attract* workers from outside its borders. Workers were recruited (often forcibly and coercively) to work on the farms and mines that underpinned the economy. Productivity was widely understood to be limited by the scarcity of workers. This perspective on work and workers was upended in the years before the transition to democracy in ways that politics and economics continues to fail to reckon with. Instead of an economy characterized by a scarcity of labour,

South Africa's primary socioeconomic problem became widely defined as precisely the opposite: a scarcity of work and a surfeit of unemployed people (see Lawhon *et al.* 2018a).

What changed?

Not productivity: the farms and mines continue to produce, but with fewer and fewer workers employed. Mainstream economic theory supposes that service sectors will grow, and in much of the global north this has been true. Jobs in the service sectors, however, are increasingly subject to automation, another trend accelerated by the Covid-19 pandemic (Stern 2016; Yang 2018; Casselman 2021). Further, many of these newly created service sector jobs are part of the excess that would be reduced or eliminated in a world of enough (see Chapter 5).

It is true that there will be many new jobs associated with greening, but current trends suggest that the net human labour necessary to create enough for all will be less than full-time labour for the global majority.

In this context, those concerned with justice and sustainability are faced with a series of difficult questions. Without substantive changes to ongoing political economies and inequalities, increased automation will concentrate wealth and increase unemployment. Orthodox socialism, and its focus on those who work, does not have an obvious solution here, for those without work have long been derided and treated as peripheral to its core concerns (Ferguson 2019). Collectives of workers in a world in which there is less and less demand for work may well produce a different sort of dystopia, resulting in concentrated wealth and mass poverty for "surplus" people without employment.

In the face of technological displacement, policies could preclude or undo technological change, an intervention that resonates with the valorization of labour in arcadian environmentalism and economics. No doubt farming with shovels could fill the gaps in employment, keeping everyone busy. If sustainability truly required a rejection of all but the most basic of technologies, or there was no other way to justify incomes, this might be sensible.

We are unconvinced that this degree of drudgery is ecologically necessary or socially just, and in Chapter 5 turn to the question of distribution and livelihoods.

A modest world of enough will require a very different set of activities, as well as different sociotechnical configurations, than modern or arcadian worlds. Cleaning and separating materials for recycling takes more time than throwing everything together into a bin. Sustainability means sometimes sewing the holes in our clothes and fixing broken toys. Some of these activities might become new jobs. For many others, however, attunement with surrounding socioecological fluxes and flows is necessary to be able to understand what to do and when to do it.

For example, modern centralized grids that dominate the provision of services in the global north are widely unsustainable: urban sanitation systems based on sewers typically flush faecal matter towards the urban periphery, a profoundly unsustainable mode of providing sanitation that wastes enormous quantities of water and treats the nutrients in faecal matter as waste. Sustainable sanitation means reworking these flows towards a more circular configuration that manages sewage locally, returning nutrients to the soil. Such configurations, however, require more human attentiveness and activity to make them work. Maintaining a biotoilet in which worms decompose material, for example, means attending to temperature, humidity and how much waste is made available; insufficient attention can mean that the worms die and waste builds up (Nakyagaba *et al.* in press). This attunement takes time, time few of us with full-time jobs are able to find in the world we have. The associated work is hard to measure and quantify, and in short fits poorly within the idea of modern, contracted, calculable waged labour. Some infrastructural labour may well be paid, but the extra time is difficult to calculate and compensate (see Stokes & Lawhon, in press; Alda-Vidal *et al.*, in press).

Modest technologies, then, are underpinned by starkly different relations between knowledge, technology and labour than either modern or arcadian ones. It is hard to imagine all the work that enables emplaced sustainability being turned into measurable, calculable labour that can be paid a just wage or equitably distributed through a collective. Instead, *Enough!* engages with these questions from a different perspective. Instead of worrying about how to keep everyone busy and find enough jobs for all, or how to justly compensate people for the activities they do that enable emplaced sustainability, it focuses on how to ensure enough for all and time to undertake the work of enough. Questions about how to distribute and pay for labour will be ongoing, but easier to justly navigate in a world of enough, a world in which there is time for democracy and doing the work of sustainability.

Ownership

Orthodox socialist theory usefully shows the limits of profit-oriented investment that seeks to maximize capital. As we hope to have made clear by now, we agree that this is a significant cause of socioecological crises. Knowing what to do instead, however, is less clear. Many anti-capitalist arguments conflate markets, private ownership and the free flow of capital, distinctions important for our analysis here. Embedding economic relations within a framework of enough opens possibilities for a more just and sustainable mixed economy that radically limits and potentially eliminates the free flow of capital. This is not the same as precluding private ownership or disavowing markets.

A modest economy does not provide, nor advocate for, a single answer to the question of who ought to own property. Aligned with heterodox anti-capitalists such as J. K. Gibson-Graham (1996, 2006) and Erik Olin Wright (2019), we emphasize that economies today are already mixes of private, state and collective ownership, underpinned by a variety of economic and non-economic logics. Heterogeneity alone, however, is not the final answer; of course already being mixed does not mean we have achieved a good economy! Contemporary economies are deeply out of balance and require extensive reworking in order to enable just and sustainable outcomes.

Maintaining economic diversity is particularly crucial given the imperfection of all responses. As we detail in the next chapter, the outcome we seek is distinct from an orthodox vision of socialist politics advanced through capturing the state to socialize ownership of the means of production. Collectives are not always just. Appropriating land for common use, for example, can create new forms of injustice for the Indigenous populations already present on those territories (Lowman & Barker 2015; Steinman 2016; Fortier 2017). Conversely, the constitution of national boundaries and exclusive collectivities can create forms of ethnic and migrant injustice (Wimmer 2002; Heimstra 2019; Walia 2021). Ensuring the possibility of existing outside collective norms might well enable individuals to push for change within or beyond established collectives. Maintaining the possibility of non-collective work also means that those whose needs and desires might not align well with existing norms still have the opportunity to participate in productive activities.

A modest approach instead embraces difference and diversity without providing a singular vision of what a modest economy looks like. A just and sustainable mix of ownership may well vary tremendously across places and sectors of the economy. Different values and experiences with collectivity, and different ecologies, will shape how much working together makes sense in particular places. For some do, and likely always will, prefer the autonomy of self-employment, even if such work is entangled, subject to socially agreed upon rules and taxation (Comaroff & Comaroff 1991; Scott 2012). Crucially, as many critics of capitalism and socialism have noted, enterprises run by individuals and collectives operate under a different set of logics than firms that are subject to the profit-focus of investors. This is not to say that such enterprises will be disinterested in profits, but instead that they are more readily embedded in socioecological relations (Corwin 2018; Corwin & Gidwani 2021). Accepting social, ecological and historical differences, rather than trying to project singular universals, is crucial to a modest approach.

A diverse economy, thus, insists on the scope, in some places for certain economic sectors, for family-owned bakeries, individually owned farms and self-employed plumbers. Others, even working in the same sectors and places, might well prefer collective endeavours. There are different scales, efficiencies and risks associated with cooperation, and there will be trade-offs and decisions that shape the prevalence of these. Crucially, this is true in the world we already have! In this sense, a modest approach is not about *upending* existing economic diversity, but about ensuring opportunity to justly participate in private and collective industry. What underpins a modest approach is the democratization of the economy, making access, ownership and decision-making more just through a balance of embedded, regulated private ownership, state ownership and collective ownership.

Further, not all parts of the economy are the same: different activities and infrastructures are differently suited to particular regimes of ownership. Natural monopolies, including much of the infrastructure in the world we have, are often well suited to state ownership. Subjecting infrastructure to the free flow of capital through financialization and privatization has in many cases been disastrous given the uncertainties of knowledge and lack of choice: it is not always clear whether water is safe to drink, and one cannot choose whose water comes through a tap. Technological platforms are another useful example here: platforms like Amazon and Uber that enable participation by small entrepreneurs are not inherently problematic, but made problematic by profit-oriented monopolies (Srnicek 2017; van Dijck *et al.* 2018; Zuboff 2019). Public ownership of such platforms may well instead offer more just access to goods and services (McLaren & Agyeman 2017).

Not all infrastructures are necessarily well suited to state ownership, however. While state ownership of modern grids is likely to be preferable to privatization, a modest approach to technology means reworking and potentially rejecting the grid itself (see above). Smaller and more circular material flows that locally compost faecal matter are poorly suited to state ownership and governance. They are dynamic configurations that require more regular attention, shaped by myriad conditions such as unpredictable rates of filling (and thus emptying), unexpected rainfall and pests. Responsiveness is better enabled by local, collective ownership and multi-scalar governance in which day-to-day decisions are made by users and those who interact with the sanitation configuration in mundane ways (see Lawhon *et al.* 2023a).

Distribution

Growth has long been the central answer through which mainstream economists suggest poverty ought to be alleviated. This answer has clearly been insufficient for creating a world of enough. Despite much attention of late, we agree with many others who suggest debates over the necessity (or not) of growth are not particularly edifying (Van den Bergh 2011). Instead, even many arguing for degrowth suggest that early quantifications of human economies and metabolisms were largely peripheral to contemporary thought and politics (Kallis 2018). Ensuring enough for all is better understood as a question of *what* ought to increase and *what* ought to be smaller in a modest future, regardless of the net outcome. We thus follow others in positioning a modest economy as "agrowth": whether the economy (at any scale) grows is largely immaterial to our argument (Van den Bergh 2011). The analytical distinction that is important is that a just and sustainable economy must be one that does not require growth, is not motivated by and centred around growth.

If the answer to enough for all is not to be found through expanding the economy and its outputs, how is it to be ensured? Further, in a world in which there is enough in a general sense, but *some* goods are *scarce*, how are they to be fairly distributed? Even in a world that pursues growth, spatial, material and ecological constraints mean that there is not an infinite supply nor always available equivalent substitutions (Meadows *et al.* 1972; Pierce 2022). There are not enough matsutake for everyone; not everyone can attend the World Cup finals. Moreover, not everyone wants to eat mushrooms nor attend football matches: few would defend proposals for equal distribution of all goods and services!

Money has long been a central answer to this question. While monetization is often considered neoliberal, money itself, of course, long predates capitalism and it is important not to conflate the two. Money is a fungible mechanism through which to enable the expression of preferences. This fungibility is crucial in a world of difference, a world where there is enough for all, but not an equivalence of all things or desires. In this context, even most anti-capitalists agree there is a need for mechanisms of exchange and expression of value preferences (Marangos 2004; Hahnel & Wright 2016; Wright 2019). Markets, we agree, are certainly always imperfect, and always will be. A modest approach is not a search for a perfect answer, but instead for the most just of available options in the world we have.

Markets have become more problematic through their excessive disembedding from the social and material world. Focusing on the struggles over market relations, Polanyi (2001 [1944]) stressed that the liberal market was necessarily a planned development. He recognized that the stability of capitalism relied upon the adoption of forms of collectivism and planning to ensure an organized social life alongside the market. While capital is inclined to reduce the meaning of human freedom to simply the ability to trade assets and sell one's labour, Polanyi stresses there is an inherent double-movement whereby this tendency is offset by the social demands of the public. Thus, markets are continuously remade through liberal capitalist movements to disembed the market from social regulation and counter-movements that resist and place limits upon the commodification of land, labour and life.

Since the 1970s, however, there has been a profound movement towards disembedding markets from social and ecological context without an equivalent counter-movement (Vosko 2000; Harvey 2005). The modest approach we develop throughout this book is rooted not just in demanding that economies be embedded: it works to articulate strategies that *enable* the embedding of markets in social, ecological and political relations through a range of already-existing mechanisms such as increased participation and economic distribution through cash transfers. Putting these already-existing tools together as part of a wider modest approach enables us to see these not as ameliorative changes, but as working towards a progressive vision of an embedded market.

Money and markets, thus, are an imperfect way to allocate goods, but are a necessary part of a modest politics of enough for all. They may cease to be the primary way some people access some goods and services, but they have a role to play in ensuring access even for those who do not conform to collective norms. In this context, a redistribution of money is crucial to entangled autonomy, and a modest economy.

Further, even if enough could, quantitatively, be assured in a highly unequal world, a more equal distribution would still be essential to a modest approach. An equitable economy is necessary for real political democracy. In this context, there is a need to both undermine structures of profit-making and disaggregate historical accumulations. Redistribution is, therefore, important not only to ensure enough, but to enable the wider political vision we have developed here.

In our consideration of redistribution, we focus on taxation as one example of a state-based mechanism through which to achieve redistribution. Taxation is undoubtedly a rather mundane approach, a longstanding one with seemingly deflated radical potential.

We mean: we know, it is not a particularly *exciting* answer.

Yet a reinvigorated politics of taxation is a key and underutilized strategy through which to pursue economic justice (Dietz & O'Neill 2013; Bregman 2020b). Taxing the rich as well as corporations is widely popular, and a truly democratic world would produce a very different tax regime (Coelho 2019). Changing who is taxed, and how much, is not easy and already there is longstanding, entrenched resistance. Anger at the low tax rates on elites may be reaching new highs as there is increased attention to the billionaires who profited from the Covid-19 pandemic (Gneiting *et al.* 2020).

Breaking the link between elites and the state is not an easy process, and we recognize that there is much beyond the scope of this book that is necessary in order to undergo this change. Yet the many pieces we have argued for here are meant to, together, increase the ability of the public to successfully make demands of the state.

A key element of rethinking taxation is to return to the question of the state. The past 40 years of neoliberalism has shaken many people's belief in the power and possibilities of the state. As we detail in the next chapter, much state action has been actively harmful and we cautiously note some congruence with the critique of modern statecraft found in some conservative arguments.

But we fundamentally disagree with the conclusion of conservative approaches to the state: that some state actions have been problematic does not mean all state action is problematic. While terminology only gets us so far, having language that distinguishes different types of statecraft is a first step towards reimagining and enabling demands for just statecraft. Granting the state power to increase taxes and redistribute wealth, as we argue in the next chapter, does not actually have to increase its biopolitical power!

Conclusion

In contrast to much political economic writing, this chapter frames historical and ongoing debates over capitalism, socialism and alternative economies in relation to modernity. In doing so, we demonstrate the limits of both modern and arcadian solutions, and the norms and assumptions that underpin them. We highlight that pathways to change within modern and arcadian economic visions may well exacerbate existing inequalities that punctuate our world. We also seek to recover another tradition of economic practice, often obscured, that reflects aspirations for a modest economy. Drawing upon the diversity of heterodox economic thinkers, we centralize longstanding practices and demands for redistribution (from gifting societies to anti-feudalists to Catholic distributionists to basic income advocates).

We then advance the idea of a modest economy that begins with the premise of enough for all and consider how thinking through sufficiency can inform our collective approaches to technology, ownership and distribution. Instead of unbending faith in scientific advance or fearful rejections of technology, we propose a politics that both recognizes the disruptive effects of technological change and advances a selective embrace of the modest potential for technological changes to enable time for the attunements necessary for emplaced sustainability and democracy. Approaching ownership, we emphasize the importance of a mixed economy with a diversity of ownership structures, ranging from neighbourhood associations to small private ventures. Economies will vary across places and time, and this diversity is part of bolstering malleability, autonomy and equity.

Despite much fixation on the term of late, our argument is neither for nor against growth; instead, it starts from the premise that enough is not a fixed quantity but possesses variable qualities associated with different political ecological decisions over how we organize the economy. Finally, we stress the need for a modest economy premised on a distribution that ensures everyone has enough. Enabling such an economy requires consideration of politics, and the kinds of demands we might make on the state, and it is to such questions that we now turn.

4

A MODEST STATE

The scale of ongoing crises can be overwhelming: there is so much that needs to change. In this context, many have looked to the state as the key agent to induce and guide such change. Mainstream politics focuses on recruiting voters, winning elections and passing laws. Battles between political parties are central in the media in many parts of the world, underpinned by emotive language in increasingly polarized discourse.

Those of us concerned with justice and sustainability are often focused on grabbing the reins of power and directing state action.

There is, however, a long line of political thought sceptical of the possibility of pursuing justice through the state. Such literatures emphasize that state power has often reinforced inequality and suppressed difference. Whether and how the state actually *can* or *would* be able to act justly remains a matter of much debate and hinges, in part, on what is meant by justice. In this context, activists and scholars have questioned whether to work with, against or beyond the state (see Rose & Miller 2010; Pierce & Williams 2016; Pellow 2017; in political ecology, see Angel & Loftus 2019; Loftus 2020).

It can be difficult, in the world we have, with our ongoing sense of urgency, to step back and think about what the state is and is not well equipped to do. Our argument in this chapter is not that state power is unimportant: we believe it is. Our central contention is that we collectively need to think more carefully about *what* we demand of the state. What we demand must be informed by careful consideration of what the state is (and is not) able to do justly.

In keeping with the wider framing of this book, in what follows we first review modern ways of thinking about the state. In short, modern statecraft is underpinned by modern ideas of justice, and modern approaches to redressing sustainability. It means rational management through measurement, calculation and a universalized vision of equality. It is a reasonably coherent narrative about understanding the world, taking power and wielding it for good.

We are, again, sympathetic to those who believe in this story, but concerned that it maps poorly onto the world we have.

As postcolonial, feminist, Indigenous, anti-racist, anarchist, environmental and other bodies of scholarship have shown, modern statecraft is based on problematic assumptions around knowledge, control and a universalizing vision of justice. Even if well-intentioned people took power, justice and sustainability cannot be achieved through modern statecraft. In a world of difference, justice cannot be fulfilled by states who directly provide to address universal, knowable needs. In a world of uncertainty, sustainability cannot be met through rational state management. Justice and sustainability both require more intimate forms of knowledge and decision-making.

Instead of *capturing* state power, some scholars and activists argue for *subverting* state power and seeking change outside the state (Day 2006; Holloway 2002).

This framework of modernity and its rejection is useful for understanding the contours of ongoing conversations. It helps us make sense of why some embrace the state as a site of struggle while others reject the possibility of just statecraft. Modern statecraft is, however, not the only form of statecraft. It is possible to agree with critiques of modernity and still consider the state as an important site of change. In what follows, we develop a modest approach to the state that moves beyond questions of whether the state ought to do more or less. Instead, we emphasize a qualitative difference, considering *how* the state might act in ways that do not universalize populations and centralize authority. This modest approach would change many aspects of what we collectively demand of the state, but in keeping with the wider focus of this book, we focus on the state's role regarding sustainability and loosely defined material welfare.

A modest approach will not spontaneously emerge within the state but instead be driven by a shift in what is demanded of the state. Our theory of change is not based on teaching state officials and politicians to practise modest statecraft; our core audience (at this stage) is not the state and its police and bureaucrats. Instead, we suggest modest statecraft might be – and in some cases, is already being – produced through shifting demands of citizens and social movements. In other words, we seek to convince those concerned with justice and sustainability to more deeply consider what actions to demand of the state, and to demand those that can harness certain forms of power without expanding its role in determining the content of a good life.

Infrastructure and UBI continue to provide different but complementary anchors for our argument. Modest infrastructure not only reworks ideas of sustainability and metabolic flows, but also the roles and responsibilities of different actors in infrastructure configurations. Through examples of infrastructure, we show how a modest state can provide oversight and facilitate just and sustainable relations without needing to know and control nature or citizens. This modest infrastructure – and more broadly, our modest political ecology – is undoubtedly more time-consuming for users than modern infrastructure. Modest politics requires freeing up time for democracy and doing the work of emplaced sustainability. We position a UBI as a unique mechanism through which to not only redistribute the outputs of the economy (Chapter 3) but also to enable time and resources without dictating how these ought to be used (see Chapter 5).

The wider implications of our argument for modest statecraft, however, go beyond the governance of infrastructure or income or any specific policy. A modest approach is a reworking of subjectivities and expectations of environment-state-citizen-economy relationships, of recognizing possibilities and limits. It is, fundamentally, a shift in understanding, including our understanding of what to demand from the state.

Modern visions of the state

The state is widely seen as a site of political struggle, a site of power that can be wielded in different ways. Capitalism and socialism, as described in Chapter 3, are both underpinned by modernity, but differently envision the role of a modern state. Socialists and social democrats tend to share the notion that the state is an institution capable of organizing social life towards justice and sustainability, an idea that underpins modern approaches to statecraft. The abuses of twentieth-century socialisms instigated extensive reflection on the possibilities of state power. In this context, scholars have, generally, moved away from established dogmas about the revolutionary capture of the state as the means to transition to a final just society (see Box 4).

Instead of approaching the state as a singular body to capture, socialists increasingly consider it contested terrain where movements can strategically advance legislative reforms and policy shifts (Poulantzas 1978; Jessop 2007). Social democrats and liberals, too, including many feminists and anti-racists, treat the state as a site of strategic struggle and compromise (Kymlicka 1995; Alstott 2004; Cudd 2006). State power is seen not as a singular capacity that can be possessed, either by capitalist elites or their socialist adversaries. Instead, the state, as well as the broader political terrain, is understood as a complex field of relations through which power flows, mediating between dominant interests and counter-movements fighting to ameliorate oppressive conditions.

To achieve their aims, advocates of justice and sustainability strategically engage different constituencies to elect officials, reform legislation, change policies, and shift state practice towards justice. Thus, although there is widespread recognition in social movements and theory that states still privilege established interests, slow processes of reform and extension of the state apparatuses are imagined to offer a democratic path towards justice. This vision for capturing, guiding and expanding state power underpins modern approaches to social welfare and environmental sustainability.

Modern states and material welfare

The role of the modern state in ensuring material welfare has largely been a politics of "more", focused on a progressive movement of the state towards more expansive labour regulations and direct provision of basic needs. Advocates of economic justice have long sought to empower the working class through a variety of mechanisms, from regulating the length of the working day and minimum wage to empowering unions and worker collectives. Ambitions include decreasing rates of labour exploitation, increasing worker control over the labour process and extending public ownership of the means of production. Associated initiatives have simultaneously sought to regulate the excesses of the private sector while expanding the domain of public enterprise and ensuring full employment.

In this section, we focus primarily on considering the role of modern statecraft in social reproduction, or providing for basic needs (we consider the politics of productive labour in Chapter 5). Socialist and feminist movements have long worked to improve conditions beyond the sphere of production, drawing attention to struggles over welfare on an everyday and generational basis. As feminists have long stressed, the capitalist economy relies upon unpaid labour, typically undertaken by women, in the sphere of social reproduction in the home (Dalla Costa & James 1972; Federici 2004; Bhattacharya & Vogel 2017). Social services, such as education and health care, also play a key role in maintaining and reproducing the population. The modern politics of more is rooted in a demand for increased and improved state provisioning of goods and services.

Schools, for example, enable families and communities to reproduce forms of social capital that structure stark differences in access to educational opportunities based on social and geographic differences (McCreary et al. 2013; Nguyen et al. 2017; Cohen 2021). Similarly, access to health care, and the debt loads accumulated and premature deaths experienced within families that lack the resources to access medical services, significantly mediates life chances. Struggles over public space and housing include anti-poverty and anti-gentrification movements that seek to control processes of urban land development and ensure that everyone has a right to a home (Smith 1996; Mitchell 2003; Purcell 2014).

Moreover, the expansive critical literature on infrastructures, including water, waste, electricity, transportation and telecommunications, raises questions about the material arrangements that underpin social reproduction and participation in public life. Modern infrastructure is founded on the idea that services ought to be provided through large-scale technological systems run by experts and subsidized, owned and governed by the state. While the promise of the modern infrastructure ideal was that all citizens would have access to services, this promise "splintered" in the 1980s as privatization and financialization shifted ownership, costs and principles of access (Graham & Marvin 2001; see more recent discussions on finance and governance in O'Brien et al. 2019; Ponder 2021). Of course, in the global south, modern infrastructure was never available to all citizens, yet this "ideal" (that someday networked infrastructure systems would be widely available) underpinned postcolonial modern politics. Both social movements and critical scholars have advanced critiques of the privatization of social services and infrastructures, often urging a return to modernist visions of state ownership and subsidization and even, for some, state provision of free basic services (largely implicit in Graham & Marvin 2001, and more explicit in texts such as Morales *et al.* 2014; Bond 2019; Satterthwaite *et al.* 2019).

Universal basic services (UBS) might be understood as the most expansive articulation of the state provision of the material foundations for life (Coote & Percy 2020). The phrase is of recent genesis, entering academic discourse in 2017 (Portes *et al.* 2017). It explicitly draws on the experiences of many countries where states directly provide some basic goods and services for free. Health care and education are the most well-known examples, and the authors of this foundational text suggest that the justifications and lessons learned from such areas can be applied more widely towards a vision of providing all basic services for free to all citizens. The term "services" is used broadly to conceptualize the range of basic needs, from conventionally understood services like schools and water to shelter, sustenance and legal aid. Sustainability is not central to most articulations of UBS, but it has been suggested that increased state control over services could spur the shift towards sustainability (e.g. Büchs 2021). We consider UBS in more detail later in the chapter.

In short, whether focused on labour relations or welfare, those working in this tradition see the state as a site of contestation and state power as something to capture and wield for social good. They demand that the state owns more and/or provides more to those with less, often at non-market rates. The modes of provisioning, too, are modern: this is most clear for infrastructure in which provisioning happens through modern sociotechnical systems. The operation of such infrastructure, and the volumes of what is distributed, are based on a rational calculation of needs intertwined with appropriate behaviours, to be provided to citizens based on a negotiated and contested but ultimately universalizable metric of worthiness. In our review of the limits of this approach below, we provide more concrete examples and critiques of modern provisioning by the state.

Modern states and the environment

Environmental actions are undertaken by modern states in many different ways, both directly reworking socioecologies and regulating the activities of others. Modern states seek to control environments through rational interventions, making them more legible, predictable and productive. This approach is part of what has been called ecological modernization; here we continue our explanation from Chapter 2 with further examples.

Responding to the arcadian tendencies of environmental politics, in 2004 a faction of the ecomodernist camp aggressively and directly attacked the mainstream environmental movement, blaming its purported commitment to a politics of less for the failure to implement more sustainable policies and win public support for environmental causes. *The Death of Environmentalism* (2004), a widely shared critique of environmentalism by long-time activists Shellenberger and Nordhaus, challenged the story of sacrifice that pervaded the movement. The authors claimed that the vision of a future of limits, less and constraints was at the heart of the failings of the environmental movement, and that, instead, science and technology could solve environmental problems without needing to rein in consumption (see also Hajer 1995; Asafu-Adjaye *et al.* 2015; Symons 2019). Crucially, while sometimes considered "outside" of environmentalism, there is a long history to the idea that science and technology can be used to redress environmental concerns.

As we noted more briefly in Chapter 1, imperial science had long sought to control unruly natures (Grove 1995). In the United States, Theodore Roosevelt, widely celebrated for his role as president in extending US federal protection over lands and wildlife, espoused a modernist vision of natural resource conservation. In his approach, he was informed by the work of Pinchot, the first chief of the US Forest Service. Pinchot is often cast as the utilitarian foil against forest preservationist and Sierra Club founder Muir, who was deeply opposed to commercializing nature. Yet notably, Pinchot attributed the deforestation of the US American West to the greed and short-sightedness of private capital, and called for the state ownership of land, a perspective that accords with much contemporary socialist thought. In his view, public lands should be managed by scientists who would collect data, monitor and make decisions to ensure the long-term production of materials for human consumption (Worster 2016).

This optimism about the potential for state control of the environment is often considered to have its roots in the Enlightenment, and expressed in the scientific management practices of Europe and entangled with colonialism (Livingstone & Withers 1999; Smith 1999).

As the practice of modern forest management developed and spread, the links between state and private capital varied: many forests became owned by the state, put into collective trust to ensure long-term sustainability. Private companies could then take resources from this land, with the state overseeing such practices. Of course, in practice, even state-owned forests have often not been rationally and sustainably managed: pressures from companies to increase rates of exploitation have resulted in widely controversial relationships between state ownership and private profit (Prudham 2007), as well as enduring concerns about the dispossession of Indigenous peoples that historically occupied the forest (Kosek 2006; Tindall *et al.* 2013).

While capitalist extraction regimes demonstrated the folly of development agendas focused only on immediate returns, the scientific and bureaucratic hubris of centrally planned socialist economies created some of the great environmental tragedies of the twentieth century. In Tanzania, the ujamaa programme of agricultural collectivization, while boasting some initial success, ultimately led to much harm. The state used universal models that forced resistant farmers to collectivize, distorted markets in ways that contributed to a decline in agricultural production, failed to not account for variance in rainfall and decimated ecological diversity in an effort to maximize crop yields (Shao 1986). Similarly, the proposed agricultural revolution in China sought to maximize food production but became a massive social and ecological catastrophe (Shapiro 2001). Following the slogan *ren ding sheng tian* (man must conquer nature), the Maoist central government sought to subjugate nature to human will. Massive campaigns of deforestation and agricultural collectivization, as well as gigantic projects for hydropower, resource extraction, and industrial development, devastated local ecologies. Yet Mao's conceit – "to use natural sciences to understand, overcome, and change nature" – echoed those of Roosevelt in an earlier generation (Mao, as translated by Ho 2003: 37). As Peter Ho (2003) argues, Chinese scientists aimed not simply to exploit the world, but to do so efficiently and rationally so as to chart a course for development that would support future generations. The tragedy of socialist planning was not disregard for the future but a more fundamental inability of government authorities to effectively direct the scientific management of natural resources in the present.

In addition to the ownership of land and forests, the state also became a central actor in the development of infrastructures that sought to control environments and provide resources. Big dams, for example, sought to rationally manage water, shifting the "wasteful" natural flows to more efficient distributions (Mwangi 2007; Swyngedouw 2015; Gergan & McCreary 2022). For instance, in Canada, the provincial government of British Columbia dammed the Peace River in the 1960s. Flooding 350,000 acres to form Williston Lake, the project created the third largest artificial lake in North America. At the time of its construction, the powerhouse was the largest of its kind world-wide and still provides a third of provincial electricity. The project, governed by a newly formed public utility, also occasioned a substantial injustice for the Indigenous residents of Tsay Keh Dene and Kwadacha First Nation who inhabited the river shores and lost their homes and autonomy in the wake of project flooding (Cox 2018).

The state, in sum, has long been a key actor in owning resources and building infrastructure to control and make unruly natures more "efficient". The mainstream environmental movement that emerged in Europe and North America in the 1960s, however, made rather different demands of the state in the context of rather different ecological concerns. Amid the Cold War, there was little demand for taking over the polluting industries, and activists instead largely called on the state to pass laws that would restrict harmful activities. (Although elsewhere, of course, states already owned such industries and were the subject of environmental concerns, e.g. Gille 2007.) As a result, in places where industries were privately controlled, a plethora of what subsequently became known as "command-and-control" regulations were adopted. Not all such regulations sit neatly within contemporary discourses of ecological modernization, but the broad idea of command-and-control regulations is underpinned by the idea of modern statecraft in which a rational state knows and controls.

Together, in short, a modernist approach to sustainability was founded on efforts to increase the role of the state as owner, developer and/or regulator. It is premised on the assumption that the state has sufficient knowledge – and that nature is sufficiently knowable – to develop efficient systems of resource generation and regulate harm. Science and technology continue to be under-stood as central to this process, enabling the production and wise use of resources and optimized reductions of environmental harm. As we noted in Chapter 1, some have critiqued ecomodernism in practice for its close asso-ciations with capitalism. Yet as we also detailed in Chapter 3, much socialist thought too has long been associated with modernity. In this context, some have argued for a socialist ecomodernism in which a strong state guides sus-tainability. Here, much as the vision of socializing modernity is underpinned by the ideal of a rational state able to effectively define the universal rights and needs of citizens, the dream of an ecomodernist state is underpinned by the idea that technoscience can effectively evaluate and regulate human–environment relations.

The limits of the modern state

Modern statecraft has produced many positive effects, and many continue to believe in it as a solution to the many ongoing crises in our world. As we have noted before, modernity is a compelling imaginary!

While some continue to embrace and demand better and more just inclusion in modernity, others have shown that this modern approach has and continues to produce enormous injustices, inequalities and unsustainability. Plantations and big dams, the relocations of people through the creation of protected areas, and a plethora of examples of "development" that integrated people into unjust economic relations *have not created a modern world as envisioned by modernity* (see Box 2). Instead, modern statecraft continues to fail for the same reasons as modernity more generally: nature and people are not uniform, knowable and predictable. There is no equilibrium to be maintained, no uniform citizen that can be provided for. Rather, within a dynamic world, there is a continually emerging heterogeneity of beings, both human and more-than-human, with diverse needs and drives (Grosz 2004, 2011; Braidotti 2013). Given this, modern statecraft is a radically inadequate tool for the promotion of justice and sustainability and regularly creates and exacerbates social and ecological harm.

Concerns with modern statecraft became more central within the Western academy through the work of critical intellectuals such as Michel Foucault. He showed how biopolitical regimes contained and marginalized populations deemed insane, criminal, unhealthy, irresponsible or sexually deviant (Foucault 1972, 1977). Power, in this view, is not simply concentrated in the hands of the few, as a dominant class or political elite, but extends through a capillary network beyond the state, shaping broad social discourses. These discourses provide us with social norms about how we ought to live, who is deserving and who is delinquent.

Much of Foucault's work and those who draw on his ideas is genealogical, showing these new forms of power and explaining their emergence and operation (Rose 1999). Much of the anti-kyriarchal scholarship that has drawn on Foucault has been more explicit in its critique. Extending this analysis to encompass broader concerns with race, gender and sexuality, scholars have persuasively demonstrated that modern statecraft relies on the normalized exclusion, abandonment and targeted destruction of particular populations (Weheliye 2014; Puar 2017; Mbembe 2019). Specifically, they show that the creation of modern citizen-subjects of state power continually relies on the projection and universalization of particular experiences and perspectives, typically those of a white, heterosexual, able-bodied, male ideal. For example, contemporary welfare and development policies have been shown to deploy biopolitical power to change behaviours and subjectivities in order to create "good" subjects

(Sigley 2006; Li 2007; Kipnis 2008; Palmer & Winiger 2019). We elaborate on this point further below, where we consider the relative merits of universal basic income and universal basic services.

The state, then, is not just a site which needs to be captured by a more diverse set of interests: even a well-intentioned state that seeks to work in the interests of the collective is oppressive when it reproduces social norms that define who citizens are and what they ought to be. These norms are not politically neutral, but generated in power-laden contexts in ways that universalize and exclude or punish those who do not conform.

Writing largely in parallel to these conversations around social and economic justice have been critiques about the hubris and exclusions associated with modern environmental statecraft. The unintended consequences of modernist state interventions to control nature through reworking socioecologies have been pointed to by many critics. Agricultural intensification produces soil erosion and increasing reliance on artificial fertilizers; nuclear accidents showcase the dystopian potential immanent to failed technopian designs; scientifically managed rangelands grow increasingly unproductive; plantation forest yields decline; the concentration of urban wastes create cascading environmental management problems; and, of course, the unanticipated effects of industrialization and auto-centred urbanization are producing our ongoing climate crisis. As we wrote in Chapters 1 and 2, modern science has been widely shown to be unable to produce the kinds of knowledge that enable people to predict and control nature. Many across the political left and right have thus concluded that the state is ultimately incapable of planning socioecological change as well as developing and enforcing rules that are deemed fair, sufficient and enforceable.

Neoliberal statecraft and market-based environmentalism

Critiques of modern statecraft have demonstrated the limits of the science, knowledge and the interpretation of justice on which it relies, as well as the injustices and unsustainability it produces. In this context, neoliberal and new conservative populism have become predominant frameworks forwarded to address the shortcomings of modern statecraft. Neoliberal environmental solutions gained ascendancy as policy prescriptions to address stagnating rates of profit in the 1970s (Harvey 2005). The neoliberal solution proposed to simultaneously ameliorate subaltern demands for liberation and elite concerns with declining profits advanced a new universal ideal of market freedom. Eliding questions of land and labour, in some places a new political consensus emerged around the importance of consumer choice.

In the global south, such ideas spread not through political consensus but through neocolonial relations. Formal decolonization was followed by development initiatives that proposed to integrate former colonies into a new international order. In the late twentieth century, policy prescriptions imposed through international financial institutions led postcolonial states to make structural adjustments, liberalizing policies to open up their markets for foreign investment and privatize infrastructure (Brown 1995; Reed 1996). Following the turn of the century, driven both by critiques of structural adjustment programmes and elite desires to make new spaces for capital investment legible, there has been a major push towards incorporating informal economies with the legal structures of the state, for instance, titling properties and "professionalizing" and legitimizing particular workers (Tokman 2007; Galdino *et al.* 2018). Simultaneously, financial inclusion has been forwarded as the means to address historical marginalization, extending a network of microloans to advance a program of private debt-backed development (Schreiner 2001; Roy 2010).

Within the global north, neoliberalization has radically transformed advanced welfare states, which had already developed substantial capacity to surveil the population and assess those who needed and deserved public support, as well as who should be excluded from the community (Soss *et al.* 2011). Large state-run asylums were abandoned and, in an era of deinstitutionalization, the insane were released to inadequately resourced regimes of community care with their families, and more punitive approaches to social deviance were adopted (Whitaker 2001; Parsons 2018). Workfare policies began to supplant welfare policies, requiring able-bodied recipients of aid to perform menial and often degrading labour in exchange for state support (Jessop 1994; Peck 2001). Immigration policies shifted from explicitly racial frameworks to increasingly economic considerations, with the rise of new temporary migrant worker programmes and increasing precarity of undocumented migrant workers (Sharma 2006; Walia 2021). New educational reforms, promoting school choice and competition in funding, were proposed to deal with criticisms of school quality; they have exacerbated some old forms of inequality and inaugurated other new ones (Nguyen *et al.* 2017; Holloway & Kirby 2020). Simultaneously, there was an expansion in the use of public–private partnerships as well as outright privatizations of public infrastructure, often resulting in increased fees for service users.

While these changes have typically been presented as deregulation and privatization, it is crucial to recognize that they embody a distinct and different mode of statecraft. The state, here, does not act in the same way as in modern statecraft: it is no longer a knowledgeable guide and provider of rules or basic needs. Instead, neoliberal statecraft produces the knowledge to make the population visible and facilitates the emergence of new markets, establishing regulatory environments conducive to maximizing investor returns.

In the field of environmental regulation, market-based solutions were advanced as overcoming the limits of modernist command-and-control regulation. Such solutions are rationalized on the basis of the incapacity of the state and a valorization of the forms of knowledge that circulate through the market. However, as above, it is crucial to recognize the centrality of state intervention to processes of market creation, constructing the boundaries of defined entities to be traded and systems for communicating their value. In order to make these markets function effectively, the state has coordinated processes of defining and regulating new commodities, such as biodiversity, wetlands, endangered species habitat or atmospheric carbon (Bailey *et al.* 2011; Robertson 2006).

These programmes are rationalized on the basis that private enterprise can develop the most cost-effective and innovative ways to reduce environmental harm. For infrastructure, this meant the privatization of technological systems and the shifting of responsibility for environmental protection and system maintenance to investor-funded, profit-oriented companies. Here, environmental impacts were to become evident and calculable within economic processes, with the state continuing to provide oversight to ensure quality of services and compliance with ongoing environmental quality regulations. Yet much as older command-and-control environmental policies floundered on the incapacity of the state to effectively measure the complex interactions of ecological systems, neoliberal solutions have typically relied on the reduction of ecosystem characteristics to simply defined attributes that can be given a form of equivalence and traded on the market (Ernstson & Sörlin 2013; Lave 2012). While these new markets enable much more flexible and nimble arrangements that can actively respond to shifting economic and environmental dynamics, they fail to adequately capture ecological complexity. Instead, they have externalized myriad environmental considerations from the very markets purported to incorporate environmental issues into economic decision-making.

Market-based solutions, of course, have certainly not brought about the good, green world we seek and have been subject to extensive critique (Bailey *et al.* 2011; Robertson 2012; Ernstson & Sörlin 2013). Some ecosocialists have attributed the failings of such approaches to capitalism (see Chapter 1). However, while they rightly reject the market-based solutions of neoliberal statecraft, they fail to acknowledge the extent to which these neoliberal initiatives respond to serious and fundamental failings in modern approaches to socioecological governance. The centralized institutional apparatuses of the state were never able to adequately appraise complex socioecological dynamics and effectively design and implement solutions.

In no way do we believe that the market-based solutions that continue to strongly shape environmentalists' toolbox are sufficient – an economy dominated by capitalist relations cannot and will not ever be just and sustainable! But understanding neoliberal solutions historically means recognizing them as a response to the very real failure of modern approaches to sustainability.

In other words, the failures of neoliberal environmental statecraft do not mean we ought to return to modernist thinking in our search for solutions.

Instead, there is a need for a new approach *beyond* modernity and neo-liberalism.

To reiterate: the turn to an unfettered market to redress concerns with justice and sustainability is flawed. Yet the underlying critique on which it is based – on the need for more flexible models of governance that enable those with more direct knowledge a greater role in decision-making – is one we agree with. The neoliberal solution of the market, however, *does not actually resolve this critique*, for markets do not actually ensure that people with more direct knowledge are engaged. In fact, it often does the opposite: it subsumes the interests of those who know and rely on a particular place to powerful economic interests. Marketization constructs forms of abstract knowledge to be traded as an interchangeable commodity rather than empowering those capable of acting to ensure entangled autonomy and emplaced sustainability.

What, then, might other options entail? We believe that it is possible to promote increased control by those more directly entangled in socioecologies without relying on capitalist market-based solutions. However, localization alone, and particularly anti-state versions that simply rely on localized knowledge and community action for solutions, will not create the future we desire. In the next section, we briefly review the limits of working outside the state before turning to what modest statecraft might entail.

The limits of arcadian governance outside the state

In contrast to both modern statecraft and market-based solutions promoted by the state, some have advocated for working to promote justice and sustainability *outside* the state. Rather than attempting to grab the reins of state power, these movements operate at a distance from the state. In this sense, they invert modernist approaches, displacing the need for centralized apparatuses of state knowledge and power. We reviewed some of the limits to some versions of postcapitalist and postdevelopment economies literature in Chapter 3 and particularly their difficulty with overcoming existing economic inequalities.

Here, we add that arcadian solutions are also inadequate when it comes to environmental governance. Ecologically, many localist practices may well have small and immediate benefits, but this is not always the case. For example, when it comes to feeding already-existing populations, local food may not be better, especially if local environments cannot produce enough food without energy-intensive farming (Monbiot 2022). Further, as many before us have noted, most environmental impacts are not contained in a local container. Localisms fail to redress cross-boundary disputes and, in a world of entrenched inequalities across many scales, do little to redress necessary concerns with distribution. They do not help us understand how to stop the large corporate actions that drive much ongoing unsustainability. They do not help us stop the environmental injustice of disproportionate impacts (Lawhon & Patel 2013). While it is clear that support for alternatives to capitalism is increasingly widespread, it is less clear how we might build a just and compelling future from the world we have.

In sum, across the north and south, there is a shared concern with state interventions that construct ideal recipients and shape their provisioning accordingly. Efforts by modern states to control nature through science and technology, and reduce harm through command-and-control regulations, have been deemed insufficient and have often exacerbated ecological harm. In the world we have, we can find inspiring examples of counter-hegemonic spaces that provide ongoing experiments in the pursuit of sustainability and justice. Yet it seems hard to imagine how, in our unequal world, without the power of the state, such places might flourish and displace ongoing hegemonies and harms.

We (Tyler and Mary) therefore agree in part with postcapitalist and postdevelopment literatures on a theory of change: capitalism and modernity are far from total systems but instead have cracks and fractures in which alternatives can grow (see Box 4). This is good and should be pursued! We support those undertaking this work! Yet, as critics have noted, it is nearly impossible to imagine substantial social, political and ecological change without multi-scalar state intervention on two fronts. The first is to ensure multi-scalar coordination and regulation, particularly for ecological issues that spill across boundaries. The second is to ensure transfers of wealth from areas of surfeit to people and places with scarcity, including but not only from the north to the south.

The politics of how to successfully scale environmental governance and enact demands for redistribution are largely elided in community economy and postdevelopment literatures, and point we work to address throughout *Enough!* It remains useful, as many have written, to think about in whose interests the state acts. Additionally, we urge more analysis of the mode of power exerted by the state through particular actions. In other words, it is essential to distinguish critiques of particular forms of statecraft from critiques of the state itself. Next, therefore, we consider what it might mean to develop and demand a different mode of statecraft. By now it will not surprise the reader, we work to imagine what a modest state might entail.

What would we want from a modest state?

How, then, can we tease out a line of action that draws on the critique of modern statecraft without rejecting a role for the state in creating the good, green world we seek? We frame what follows not around conventional questions about whether the state ought to be bigger or smaller, or in whose interests it acts, but instead with qualitative questions about what is to be done and how. This enables us to distinguish between modern statecraft (deploying state power to create knowable populations and act on behalf of knowable populations) and modest statecraft (deploying state power to protect citizens, redistribute resources and enhance capabilities). The modest approach we develop here is not a purist politics, for there is no way out of the tension between autonomy and entanglement. Instead, it is a search for a framework through which justice and sustainability might be pursued in ways that limit harms while also catalysing positive change. In the next sections, we consider how a modest state might act as an overarching coordinator and regulator rather than knower and controller of nature and citizens using the example of sanitation infrastructure. We then consider a UBI as an exemplary form of modest statecraft, for it enables the redistribution of resources without enhancing the state's biopolitical power.

Modest environmental governance

A modest approach is distinct both in terms of its approach to regulations and in how it seeks to overcome existing political tensions. Situated knowledge is necessary for sustainability, yet local governance alone is insufficient given the scales at which environmental harms occur and the mobility of those who might cause harm. A modest approach to regulation is about providing a framework of rules as well as ensuring the distribution of resources to enable sustainability. In this sense, it is distinct from neoliberal devolution, for it is premised on a redistribution of decision-making, resources and power. As we have argued throughout *Enough!*, the ideas we write about here resonate with many emergent ideas and practical experiences, and our effort is to tease out a common thread rather than to propose a new grand narrative.

As but one example, we consider the fraught experiences of governance and responsibility around the Quapaw Oklahoma Tar Creek Superfund site, a notable case of devolved environmental responsibility supported by state resources. The story of Tar Creek, an area of almost 1,200 square miles of Quapaw tribal territory, is a tragedy a century in the making. From 1850 until 1950, the region was devastated by lead and zinc mining operations. The Quapaw were marginalized as decision-makers and beneficiaries of development. The result was a landscape of ruins, scarred by lead poisoning, mine waste, acid mine water and sinkholes. To address this legacy of environmental racism, government officials began to intervene in the early 1980s. Rather than command-and-control or market-based interventions, this case highlights the potential of a mode of governance rooted in a recognition of the importance of situated knowledge in conversation with experiences elsewhere: Indigenous sovereignty and self-determination meant that state officials had to work in collaboration with the Quapaw, recognizing their commitment to landscape restoration through emplaced sustainability (Nolan 2018). Yet it also shows the difficulty of doing so in our highly unequal world, in which centralized "expert" authority remains entrenched, remediation is costly and responsibilities contested. For tensions over roles and responsibilities have arisen over the distribution of authority and funding, and contestations are ongoing, as are demands for the further devolution of the process of remediation to the Indigenous community (Burnley 2017).

Distributed funds and responsibilities have been more successful for sustainable infrastructure. Solar power development, an inherently decentralized energy infrastructure, presents a technological alternative to monolithic megaprojects (although some continue to strive for mega-solar projects, these have thus far had limited success). Kwadacha First Nation, displaced and relocated following the flooding of Williston Lake in northern Canada, had for decades remained isolated from public electric infrastructure and dependent on diesel and propane generators. In 2021, a government emissions reduction initiative provided funding to the community to build solar power infrastructure to reduce their fossil fuel dependence.

Similarly, micro-hydro run-of-the-river projects generate green energy without occasioning the massive displacements and environmental destruction of huge reservoirs. Increasingly, remote Indigenous communities have been partnering in these projects to serve community needs in a sustainable way. For instance, on the Canadian West Coast, the Tla-o-qui-aht First Nation has partnered in the Haa-ak-suuk Creek Hydro development, which produces enough electricity to supply about 2,000 homes on Vancouver Island (Miller *et al.* 2019). Solar electricity is also one of the examples developed in Klein (2014) and Lewis (2015) as part of the story of "yes" (see Box 3). Yet while it is decentralized, most solar technology cannot be locally produced and falls outside what might be typically considered as part of arcadian environmentalism. Moreover, although based on science and "extracting" sunlight from nature, there is no effort here to control the nature of the sun; users must work with its predictable and unpredictable patterns. In this sense, we might think of solar energy as modest, responding to nature's dynamics in a non-local, technologically sophisticated, attuned and accepting way.

The regulation of infrastructure provides another useful lens into what modest governance might entail. Our ideas here are inspired by research on sanitation infrastructure in Kampala, Uganda and eThekwini, South Africa. Like with the example of the Quapaw Oklahoma Tar Creek remediation above, our intention is not to suggest that any of these represent an ideal case of what ought to be. Instead, they help us to see the limits of all solutions in the unequal world we have and to better understand what modest governance might entail.

Modern infrastructure entails state ownership and regulation of networked grids, and for sanitation this means pipes and sewers through which waste is flushed away. In Kampala, less than 10 per cent of the city has access to this infrastructure, a pattern established in colonial times. eThekwini's history of infrastructure inequality is equally rooted in racialized exclusion. National and municipal governments have, at various times, through different policies, imagined a future of modern sanitation for all, but this future seems increasingly distant, and in the context of changing rainfall, increasingly implausible. Many residents, therefore, have long used non-modern sanitation configurations, and modern states have long penalized residents for practising open defecation or constructing illegal latrines even when no alternatives exist (Sutherland *et al.* 2014; Lawhon *et al.* 2023a).

Some have argued for the privatization of infrastructure, including setting up profit-oriented toilets, an argument underpinned by neoliberalism. Only those who pay would, in such a system, be permitted to use the toilets. Yet sanitation is a useful example of multi-scalar entanglement in which the costs of not using a toilet spread, even to those who might have paid for sanitation. Urine and faecal sludge do not stay in place. From pointed crises like cholera outbreaks to the everyday deaths from childhood diarrhoea, from eutrophication of regional water bodies to decomposition that contributes to climate change, unsanitary conditions can cause extensive social and ecological harm. Externalities that harm those who do not cause the problem fly in the face of market-based explanations of how a fair society ought to operate. Neoliberal theory tells us that those who cause these externalities should pay, yet this requires monitoring and enforcement – as well as infrastructure and money.

In this context, municipal governments continue to insist on the need to regulate sanitation but are exploring options beyond simply penalizing those who do not use legal sanitation configurations. At times, this has entailed providing regular services outside the modern ideal, including a range of sociotechnical sanitation options. There are countless examples of states owning and operating such infrastructures. Communities, too, have come together, at times, to provide safer sanitation options through a wide range of sociotechnical configurations, ranging from private for-pay toilets to NGO-sponsored facilities.

At times, such heterogeneous configurations work well, for many residents, at least for a time.

Oftentimes, however, they do not.

The reasons for failures are widespread, including limited resources, conflicting responsibilities, unanticipated rains and floods, poor construction and unsuitable technologies. What matters for our argument here is that, on the whole, despite much well-intentioned effort, many in the global south are unable to rely on state or community-based sanitation options to provide safe and dignified infrastructure.

Some municipal governments in the global south have decided, in the context of these dynamics, to take an alternative approach to governing sanitation (Sutherland *et al.* 2014; Lawhon *et al.* 2023a, 2023b). There is a growing acceptance that a modern vision of how infrastructure ought to be is fraught: it is both sociospatially unjust and ecologically unsustainable, underpinned by extractive material flows that take from some places and dump wastes elsewhere. Rejecting the possibility that modern infrastructure can be justly and sustainably provided for all is emphatically not a rejection of the goal of ensuring that all people have access to safe and dignified infrastructure; it does not endorse outsourcing responsibility nor advocate for penalizing those with inadequate infrastructure.

Instead, municipal governments in Kampala and eThekwini have begun to frame themselves as facilitators of sanitation services. The state, here, is no longer the owner or knowledgeable rule-maker, no longer the source of expertise working to control nature and induce right behaviour from citizens. Instead, it facilitates the experimentation and construction of sanitation innovations, helping different stakeholders learn from each other and providing overarching regulations in consultation with a range of stakeholders (Sutherland *et al.* 2014; Lawhon *et al.* 2023a).

In Kampala, this has meant changing regulations to permit the flow of faecal matter outside of the sewers and beyond state-operated sludge removal trucks. This movement of faecal matter has always been ongoing, but now local entrepreneurs can obtain state permits and training from non-state actors and legally undertake this work (Nakyagaba *et al.* 2021). The new configuration is far from perfect, but a prefigurative practice working towards new more-affordable and sustainable material flows. Various stakeholders are, for example, working together to create more pliable infrastructure open to these newly legitimized actors. At present small entrepreneurs must take the sludge to a facility at the edge of the city, a distance that significantly increases the cost of providing this service. There are ongoing efforts to add sludge collection points throughout the city, an intervention that would radically cut costs and likely ensure that more of the sludge entered regulated material flows (rather than being dumped into empty spaces or waterways). Crucially, the state here has shifted away from modernist forms of statecraft (building and owning pipes, policing illegal practices). Instead, its role is shifting towards what we call modest statecraft, characterized by facilitating and enabling heterogeneous practices, relying on imperfect and situated knowledge (Lawhon *et al.* 2023a, 2023b).

Such reworkings of the politics of knowledge and governance may be most evident in places where modern infrastructure was never fully present and operational, yet recognition of the limits of modern approaches to environmental regulation and consideration of more capacious processes are increasingly widespread. What we describe above has some affinity with the shift towards greater participation in environmental governance noted in the 1990s (Bulkeley & Mol 2003; Davies 2001; see also Gandy 1999). It also resonates with some forms of experimentation that are increasingly becoming the norm: while some experimentation is rooted in the idea of developing blueprints that can be cut-and-pasted, others embrace diversity, recognizing that there is no single model for the future. Further, some modes of experimentation are rooted not only in an acceptance of biophysical uncertainty but in a growing sense that there is not a singular set of objective criteria against which "success" can be measured (Edwards & Bulkeley 2018).

What, then, differentiates a modest approach from modern statecraft, including late modern modes of reflexive governance (e.g. Voss *et al.* 2006)? It is not simply the practice of participation or experimentation, for participation and experimentation *can* both be underpinned by modernist imaginaries. A modest approach embraces participation and experimentation as a response to a particular socioecological imaginary, one in which we can no longer reasonably be guided by blueprints and singular logics. It is not a rejection of rationality and calculation, but a recognition of the fundamental indeterminacy of nature. It is founded on a need to consider insights from those most attuned to changing socioecological conditions as well as to learn from elsewhere. It is not a rejection of the importance of the role of the state in ensuring justice but a shift from justice as rule-making and provisioning towards justice as enabling.

There is no shortage of examples of this kind of practice. Yet as we have argued throughout *Enough!*, there is a need for analysis and vocabulary that enable us to distinguish modest and modern governance, and the different imaginaries that underpin them.

That said, while examples are plentiful, none of those we have provided are perfect models of modest governance: there is no such thing. Research collaborators and participants in Kampala and eThekwini suggest that those working in these emergent governance arrangements agree: this mode of governance is imperfect, ongoing, experimental. Many working within these new arrangements are searching for more inclusive, more sustainable and more just arrangements; many are also working to make these arrangements suit their vested interests. These are not apolitical new configurations, but deeply political. We can also find ongoing efforts to weave modernist forms of control into emergent configurations. For example, some sludge removal experiments required the carrying of monitoring devices, which were resisted by many workers who "accidentally" dropped these into the sludge (Nakyagaba *et al.* 2021). Such technologies need not be deployed in modernist ways, but there is an ongoing need to think through the relationships between digitalization and modernity, and ways that such modernist impulses might be rejected without denying the utility of technology-based connectivity.

One concern with emergent modest infrastructures is, parallel to our concerns raised about arcadian diverse economies, that the actual day-to-day payments for users is often higher. This is not necessarily because such configurations are more expensive than flushing toilets connected to sewers, but because states have long subsidized the outlay, maintenance and management of modern infrastructure. How are businesses and ordinary residents meant to pay for modest sanitation configurations? Direct subsidization is complicated as there is diverse ownership, a wide range of actors and technologies, more competing interests that might vie for funds. One strategy that we consider in more detail below, which is not limited to infrastructure, subtends such questions by increasing the incomes of users. This strategy does not preclude state financing, but means that residents have greater autonomy to choose between different sociotechnical options and are able to withhold payments when services are not provided, crucial elements in the ongoing experimentation with infrastructure configurations. We turn to such considerations, and the broader logic of cash transfers, below, mindful that increasing incomes and subsidizing infrastructure may be complementary strategies.

In sum, the pursuit of modest governance is not a clear process of smooth progress towards a known ideal. We can learn from other places and experiences, and indeed there are many useful examples already ongoing that we can draw on to consider more and less just and sustainable options. These help us collectively to work towards prefigurative, agonistic processes of making the path by walking it.

Provisioning through UBI as modest statecraft

In the highly unequal world we have, how might people be ensured enough? In the previous chapter, we considered what a modest economy might entail, including the importance of redistribution and markets. In this section, we show why a universal basic income – described in more detail in the Interlude – is emblematic of a modest approach to provisioning. In short, it harnesses the power of the state without increasing the biopolitical power of the state (Lawhon

& McCreary 2023). Instead of collectively deciding the material contents of the good life, a UBI provides people with money. If no other changes were to happen, this would surely not produce the good, green world we seek! Yet as we argue throughout *Enough!*, a UBI is both a clear example of modest statecraft and can be used to underwrite this wider change.

Ongoing debates between UBI and UBS can help us to tease out these differences (Gough 2021; Koch 2022; Thompson 2022; Lombardozzi & Pitts 2020). As noted above, some argue for the direct provisioning of services by the state. Such arguments build on already-existing examples of modernist state-craft, in which states decide what services are required by different populations. This includes decisions over who gets what services as well as the degree and quality of service. For example, public schooling may be free for primary and secondary, but not tertiary, education; eyewear and dental services may or may not be part of public health services, and this may be different for people of different ages and incomes. Advocates of UBS argue for a collective determination of basic services and for these services to be freely provided to everyone by the state; we consider this a modern vision of enough for all.

Engagements with UBI by advocates of UBS were initially rather hostile (e.g. Coote 2019), but recent work raises the possibility for thinking through how the state might provide a combination of basic income and basic services (Coote & Lawson 2021). A UBI and UBS are likely to have a similar material impact for some recipients, particularly those whose needs meet the normative expectations of good citizenship. Yet a UBS is based on a fundamentally modern position on the role of the state in defining the contents of the good life. Specifically, implicit in the argumentation is that the state is better able to know and provide than the individual.

Thus, while UBI and UBS may well have similar goals of ensuring enough for all they are based on *fundamentally different conceptualizations of justice, of the relationship between citizens and the state, and how sustainability is to be sought.* A UBS channels experimentation through the state rather than users. This limits incentives for people themselves to pursue alternative technologies and relations through which to build entangled autonomous lives. With energy, for example, centralization through UBS encourages a uniform mode of accessing energy, i.e. energy from a single state-owned grid. There would be little reason for communities to initiate or participate in the decentralized modes of renewable energy production noted above. If provided a basic amount for use, there are equally fewer incentives to reduce consumption; perhaps ironically, only those whose consumption exceeds basic needs would have a clear incentive to explore and undertake such innovations.

For sanitation, a UBI would also contribute to ensuring the viability of modest infrastructure. One of the key limitations of efforts to ensure everyone has access to sanitation continues to be that many people struggle to afford this service. Even simplified sanitation infrastructure uses resources – and requires work from users. How these infrastructures can be funded remains an outstanding question. State subsidies may well be useful, yet particularly in contexts where state presence is limited and contracts are difficult to enforce, money continues to be one of the most straightforward ways to increase the likelihood that infrastructure works. Waste collection is emblematic here: there are countless examples of rubbish not being collected by the state from skips. At times, residents will instead pay private enterprises on a case-by-case basis to have such waste removed (e.g. Zapata Campos & Zapata 2013; Sseviiri *et al.* 2022). More broadly, if a pay-by-the-use toilet is broken or unclean, residents can choose to go elsewhere. In a wider context in which services are irregular, having a direct financial relationship increases the likelihood that waste workers show up, that sanitation workers show up and keep the toilets clean and working.

Some might reject this as a neoliberal system, yet as we insisted in Chapter 3, money and markets predate and exist far outside of neoliberalism. More substantively, state-provided infrastructures are more likely to fail: there are no shortage of cases in which, for various reasons from vehicle failure to illness to opportunism, infrastructure provided by a distant state simply does not deliver. Even in the global north, the toilets in privately owned shops are generally more pleasant to use than publicly owned toilets.

We might provide all sorts of well-grounded critiques here in order to explain why modern state services *ought* to work, and what checks and balances would be needed to assure that the system goes according to plan. Yet in the world we have, directly linking payments to services tends to work. Many of these examples also generate embedded economies in which people know each other, and there is often forgiveness and flexibility, emotional attachments that extend beyond the transaction (Nakyagaba *et al.* 2023).

Such relationships tend to work, at least, in places where residents have sufficient incomes in order to make these payments. In this sense, providing income rather than services enables more localized decisions, innovations and adjustments; it more closely accords with a modest approach of emplaced sustainability. This is not to say that there is no role for the state, but that in many cases the state is better suited to oversee, coordinate and support than directly provide.

Crucially, the modest technologies and modest governance we have described in this chapter will require much more engagement from ordinary citizens. This has long been recognized as one of the limits of radical democracy and emplaced sustainability, as well as the kind of modest infrastructures we have pointed to in this chapter. Enabling radical democracy is not simply about creating more opportunities for participation: central to our argument throughout this book is that changes are also needed that provide people with time to participate.

Moreover, there needs to be a sharper analysis of the types of demands being made. A UBI will not magically solve any of these challenges, but instead enables people more time to spend undertaking the tasks of governance and emplaced sustainability. We pick up on this point in more detail in Chapter 5.

Conclusion

Modern and neoliberal statecraft has been the focus of much critical scholarship about the state. Many (but importantly, not all) socialists explicitly identify with a modernist lineage, relying on a strong and knowledgeable democratic state to plan and act on behalf of citizens. Arcadian approaches to the state are less prevalent, and often implicit in this literature is a desire to work beyond the state. Much postdevelopment, anarchist and community economic theory rejects the idea that modern statecraft can be progressive and eschews a role for the state in underwriting social justice. These critiques often come from writers emphasizing biopolitics and processes of subjugation, including writing about race and gender, who stress the need to stop circumscribing particular identities and recognize heterogeneity. Thus, command-and-control, capital-driven and local anti-state approaches to socioecological governance have fundamental flaws.

Yet regulation – whether by a separate institution called "the state" or agreed upon directly by communities – is necessary across spaces and scales to enact radical change. We position modest statecraft as an approach that differs from modernist, neoliberal and arcadian approaches to the state, providing a framework for understanding modest politics and modest statecraft. A modest approach draws on ideas within these literatures that do not fall neatly into the binary between modernist hubris and anti-modern scepticism (including heterodox versions of socialism, degrowth and diverse economies).

Our approach recognizes the limits of the state but also the need for the state to act in ways that protect citizens from multi-scalar ecological harm, as well as promote citizens' capabilities and economic security. A modest approach does not dictate the pathways for reducing harm. It accepts that knowledge and governance are always imperfect and uncertain; those with everyday encounters with situated ecologies are particularly well-positioned to witness, make sense of and respond to changes. Enabling change also means demanding a reallocation of resources, ensuring that people are supported in their efforts to witness, learn and relearn about their world in order to guide socioecological interventions.

In this chapter, we have considered some examples that accord with this modest approach, but emphasize that the politics of radical modesty go beyond particular governance structures or policies: a modest approach reworks expectations of environment-state-citizen-economy relationships. Working towards modesty is an ongoing process, involving a recognition that struggles are ongoing and require engaged citizens and, importantly, a socioeconomic context that enables time for doing democracy.

Some have critiqued these more fluid forms of statecraft, suggesting that they are simply a minor rebranding of neoliberalism and that they enhance the power of some actors over others. We are emphatic that, in the world we have, simply increasing participation in the state without reworking myriad forms of kyriarchal relations (not only capitalist ones) is insufficient. Even in places where most citizens want to see more environmental regulation, vested interests have made it difficult for the state to reduce harms.

There is little utility in hoping that just governance might suddenly appear.

Instead, more equal governance arrangements must be iteratively built from our fraught world through imperfect, impure prefigurative practices. It will not surprise the reader by now that we believe a UBI might be used to subtend conflicts, enabling funding and time to create and manage modest infrastructure as well as participate in modest governance. In the next chapter, we continue our explanation of why a UBI accords with, and might underwrite, a modest political ecology.

5
MODEST LIVELIHOODS

How do we get what we need to live? Many of us imagine a straightforward answer: we get a job, the job pays us money, then we use that money to buy what we need.

But for most people in the world, it is not so simple.

The notion of livelihoods was developed by practitioners working in the global south to expand our thinking about the diversity of ways people get what they need (Owusu 2020). As we imagine our way towards a world of enough, it is a useful concept to help us think beyond work-based incomes in the global north too.

Work and incomes are not stable: some food, energy and water is obtained directly from the local environment, and social relations mediate access and distribution. Even for those with a single job and income, getting what we need to live is often more complicated than going to a shop. What impacts whether a person has enough is not just about whether one has sufficient money but is shaped by the broader political economy. This, in turn, influences needs, access and prices. Infrastructure and public services are crucial considerations here: the income one needs for a basic livelihood is radically different in places with good mass transit, state health care and safe tap water. The "welfare purchasing power" (Hickel 2020) of money is higher (meaning a basic income can be lower) in a place where more goods are shared, subsidized or free.

In this chapter, we draw from this wider perspective of livelihoods to think through different ways of understanding how enough for all might be achieved. In the long arc of history, it is reasonable to imagine that the work of most people usefully contributed to basic social and material provisioning. For some time now, this has no longer been true. How much work is necessary and who does it are clearly not only – and today not even primarily – questions about productivity, about ensuring enough. They are political economic questions shaped by flawed imaginaries, socioeconomic inequalities and deeply troubling economic systems.

We continue this chapter by first considering modern approaches to, and politics of, livelihoods. This story is likely familiar, and even taken for granted as the ideal form of livelihoods for many readers. We then tease out two central values that underpin the modern work imaginary and continue to underpin politics: that there is a *moral* value to working and that this moral value means that *work* ought to be the primary source of a livelihood. We consider how and why this ideal was constructed, drawing on a long history of scholarship showing the links between modernity, colonialism, capitalism, Christianity and the shared belief in the moral virtue of work. As in previous chapters, we distinguish responses to the exclusions from modernity from the rejection of modernity.

Arcadian imaginaries again serve as a useful point of contrast, urging livelihoods from local environments, non-waged work and collectively determined direct distribution. Yet unlike in previous chapters, we find that modernist ideas have, at times, become woven through arcadian thought. Specifically, many environmentalists – including some advocates of degrowth – implicitly valorize labour beyond its use value, and adopt the assumption that work ought to be the primary source of a livelihood.

In this chapter, then, we consider modern and arcadian ideas of how livelihoods ought to be obtained. We also address a second thread significant for the politics of livelihoods, the modern idea that work creates good subjects who deserve enough. We argue that the valorization of work continues to shape, and constrain, radical politics by comparing a jobs guarantee with a UBI. We show that the modern work ethic underpins the urge to put everyone to work, even when a sustainable world of enough would be easier to achieve without the pressure to create employment for everyone.

Ultimately, this chapter demonstrates why we ought to collectively reject the modern imaginary and instead pursue a modest approach to livelihoods. The modest approach we develop here is founded on delinking basic livelihoods from labour without collectivizing material provisioning. In other words, a world of enough is premised on unconditionally ensuring enough for everyone without directly providing or defining what enough is. As noted in our discussions of technology in Chapter 3, a world of enough does not require productive labour from everyone. And, as we elucidate in this chapter, there is no moral reason to demand work from everyone. This is not to suggest no work will happen in a just and sustainable future, nor to discount that some people will likely want to undertake certain kinds of work (for supplemental income, to enable greater consumption or for the pleasure of undertaking certain tasks). Nor is it to urge delinking all work from monetary compensation; as in Chapter 3, we believe there is utility to money and embedded markets. What work is to be done and how the work that needs doing will be distributed and compensated are important questions, ones that will be answered differently across a pluriverse.

Here, we emphasize that just distributions of work are only possible in a modest world of enough, where livelihoods are not dependent on labour and therefore people are, as best as possible, freely choosing how to spend their time.

We develop our analysis of the merits of a UBI by comparing it with proposals for a jobs guarantee. We show that a jobs guarantee is underpinned by problematic modernist assumptions, including a distrust of individuals (who are presumed likely to misuse monetary assistance), a modernist valorization of work and a belief in a knowing state capable of just provisioning. A UBI, instead, opens possibilities, enabling capabilities that can be used to underwrite building new worlds without foreclosing possibilities about what will come.

The modern imaginary of work and labour politics

In this section, we explain the modern imaginary of work and how it has, and continues to, shape radical politics. The modern imaginary of work, like the wider modern imaginary, has changed over time and space, and many pages have been written tracing this history (Weber 1930; Weeks 2011). Much of the time, this modern imaginary is implicit, invisible and unquestioned. Here we provide a broad outline to emphasize how this imaginary has shaped many people's ideas of what good work is and who ought to undertake it, as well as its relationship with what it means to be a good and deserving person. Shining light on this imaginary and where it originates enables us to more clearly ask whether we (collective) really believe, and want to build a politics underpinned by, a modern work ethic.[1]

Our focus in this section is income in modern capitalist societies, and modern socialist politics within such societies. This is to complement our discussion of modern statecraft in Chapter 4 (which included consideration of how socialist states sought to ensure basic needs through collective provisioning, and contemporary proposals for universal basic services) and our discussion of a jobs guarantee below.

In one formative version of a good livelihood, a single income (from a white male labourer who had one full-time job for most of his adult life) was imagined to provide enough (for the man and his nuclear family) (Vosko 2000). Workers at Ford Motors in the United States in the middle of the twentieth century might be emblematic of this vision, but they had counterparts across the globe. The apartheid state in South Africa, for example, worked to turn this vision into reality for white men, increasing employment and wages to create a staunch middle class. In this context, advocates for the working class sought gains through participation in politics and unions, with workers making demands of both owners and the state to improve their positions (see Hill 1996; Freund 2013).

Of course, this ideal has always been recognized as a social, geographical and historical anomaly, one that relied on other, undervalued labour. Feminists, for example, have repeatedly demonstrated in countless ways that this imaginary is predicated on unpaid reproductive labour (Winders & Smith 2019; Samantroy & Nandi 2022). Anti-racist scholarship has similarly shown that the wages of this class depend on undervalued labour of non-white workers; this is particularly visible in South Africa, but is also true from Detroit to Sao Paulo (Gidwani & Chari 2004; Corwin & Gidwani 2021). Further, most work globally and historically has looked nothing like the formal, contracted, secure waged labour of the modern imaginary: from subsistence farming to family grocers, from waste reclaiming to driving a matatu, work arrangements are, and have always been, profoundly diverse, fluid and variously secure (Makhulu 2012; Munck 2013; Monteith *et al.* 2021).

Scholars and activists have pointed to such examples in order to show the racist, sexist and colonial relations embedded in formative iterations of the modern work imaginary. And yet, many critics of this formative version of the modern imaginary do not *reject* the idea of work that underpinned it. Instead, they undertake two key strategies for revising and expanding the remit of modern work: widening who counts and making non-modern work modern.

Advocates of expanding who counts as a worker sought, for example, to open union membership to non-white workers and enable women to access more parts of the labour market. They continue to work towards as-yet elusive goals of equal wages across race and gender, as well as working to change racist and sexist state policies. Others sought to demand payment for housework, insisting that reproductive labour counts (Strong-Boag 1979; see Federici 1975; Weeks 2011). Such activism has generated many benefits, and again we seek to be cautious and careful in our critique here: we (Mary and Tyler) are not opposed to more-equal work conditions and more-equal state policies. An emphasis of equity and inclusion, however, has largely left unquestioned the foundations of the modern work imaginary (particularly, its fundamental valorization of work).

A result is that the modern work imaginary has, in part through these struggles, been updated: a dual-income household of two full-time workers of any race and gender ought to provide enough for a good life for the workers and a pair of dependents.

A second prominent strategy has been to advocate for making non-modern work better accord with the modern ideal. In the global south, this has often entailed the formalization of informality, changing work to better accord with modern ideals of set hours, wages, rules and responsibilities legible to the state. For example, there have been many efforts to transform informal waste picking into formal-sector jobs (O'Hare 2019; Grant & Oteng-Ababio 2021; Luthra 2022). In practice, such transformations have often occurred in neoliberal contexts where low-waged and insecure jobs were created (Miraftab 2004). Many concerned with justice have criticized this formalization because the jobs created do not align with the modern imaginary of what a good job is, urging that just formalization requires more-secure, better-paid and unionized conditions. (Again, this is not the only strategy, and we draw below on other forms of engagement with informal work.) Critical scholarship on the gig economy, written mostly in the global north under quite different economic and employment contexts, typically parallels these concerns (Vallas & Schor 2020), arguing against the displacement of modern jobs by the gigification of the economy (Press 2022; see Woodcock & Graham 2020).

These two strategies – expanding the range of people considered to be modern workers and ensuring the conditions of labour mirror the ideal of modern work – have had many successes, and contributed to very real gains made for many across the global north and south. They have resulted in increased political recognition, notable (albeit insufficient) improvements in racial and gender inequality, and better working conditions for a large number of workers.

Many continue to believe in the promise of the modern imaginary of work and increasing the number of modern workers remains a goal for many activists and states.

There does, however, seem to be growing disillusionment across the global north and south with the possibility and desirability of modern work for all. While gains have been made in the long arc of history, the gap between the promise of the modern work imaginary and the experience of most workers is, at present, growing. Scepticism is most palpable in places where modern ideals of various sorts were always illusive: disillusionment is the wrong word in places where many never really believed in a universal modern ideal (Comaroff & Comaroff 2002). The Covid-19 pandemic prompted many people to think more about the role of work in their lives, and some propose that the ongoing "Great Resignation" is indicative of changing values and a search for alternative ways of building lives and livelihoods (Elhefnawy 2022).

What might it mean to stop believing that modern work for all is a plausible goal?

For many, it appears, disbelief and disillusionment are producing a visceral sense of crisis, of material and existential angst, particularly for youth. Some attribute growing radicalism and violence in part to un(der)employment and a disillusionment with modern political economy (Beinin 2016). In keeping with the wider argument throughout this book, and aligned with the many youth demanding not just inclusion in modernity but political change (Honwana 2012), we see the rejection of the impossibility of modernity not as just a closure: it is also an opening. Rejecting a modern work imaginary does not have to result in an acceptance of a radically unequal world, poor working conditions and insufficient livelihoods. Instead, it enables us to imagine livelihoods beyond modernity, and how they might contribute to a more just and sustainable world.

But let us not rush too quickly ahead. First, we briefly touch on arcadian ideas of livelihoods. Then, given its ongoing pervasiveness, let us spend a few more pages on the modern work imaginary, and build our argument for why it ought to be rejected.

Ambiguities in arcadia

The core of arcadian approaches to livelihoods has been explained in Chapter 3: production occurs in non-capitalist ways, and often in closer rela-tion to local economies and ecologies. This narrative has been problematized in many ways, not least of which is the struggles faced by people like Chris in our Introduction: subsistence from local environments is increasingly diffi-cult in a changing and contaminated world. Further, as we noted in Chapter 3, the theory of change in most arcadian literature provides no way to overcome existing inequalities.

Much arcadian political economic thought relies on collectively deter-mining who ought to work, and what work ought to be done (although, again, collective determination is not inherently arcadian: many canonical examples of real-world socialism and socialist thought rely on collective provisioning through modern modes of statecraft, as outlined in Chapter 4). Here, decisions over work and distribution are shared without reliance on modernist universals and norms. Yet as we have worked to show throughout *Enough!*, arcadian politics are also underpinned by social norms, and the collective decision-making that emerges is not necessarily just. There is, for example, a long history of patriarchal dominance, colonial conceits and oppression of minority communities that highlights the dangers implicit to the majority conception of the abilities and needs of others (see Young 1986; Chapter 4). There are also longstanding popular beliefs that certain kinds of work befit certain races and genders, and that welfare and consumption rates also ought to vary across race and gender (Makhulu 2016; Dicke *et al.* 2019; Thakholi 2021). A collective might well believe that women are more capable of household responsibilities, and thus decide that cooking dinner is a chore that should be performed by women, or that people in same-sex relationships ought not qualify for state assistance.

In short, arcadian versions of livelihoods generated from local environ-ments are increasingly ecologically implausible (at least, in the world we have). Collective adjudications based on arcadian values may, at times, be more just than market-based allocations, and preferable, at times, to provisioning through a modern state. There are surely examples that can be found, and principles that could guide such processes. But it is not necessarily better and can instead reinforce kyriarchal relations. Further, as we think through in more detail below, much of the contemporary arcadian political ecological and economic literature is permeated with what we have identified above as the modern work ethic, a valorization of labour for its own sake. In this context, we must unpack the origins of this valorization, and how it has shaped contemporary politics. Then, in the section that follows, we consider a proposal that has emerged and gained popularity across the political left and right, and been advocated for by both modern and arcadian environmentalists: a jobs guarantee.

Emergence and critiques of the modern work imaginary

Above, we noted that scholars and activists often seek to expand modern work, broadening definitions of who can undertake this work and making non-modern work better accord with this ideal. Here, we emphasize that this is not the only way to respond to the modern imaginary, and consider what might happen if we instead reject it.

What if, returning to the question we posed at the beginning of this book, we also considered the history and present of labour politics as a *refusal of the valorization of work?*

Again, this might seem a little strange to some readers. It was rather disconcerting to us when we (Mary and Tyler) first, not so very long ago, encountered these arguments in the literature. We were both raised to believe that working hard was a good thing, that having "a good work ethic" meant one and only one thing: working hard.

In this section, we trouble the idea that there is only one ethical approach to work, and instead label the belief that working is morally good as the modern work ethic.

For Mary, this set of questions took on particular significance in South Africa, where unemployment is widely cited as "the" pressing social issue. The question of how to address unemployment also continues to dominate conversations about economic development in northern Canada, where Tyler has undertaken much of his research. In both places, talk of job creation holds tremendous political and moral sway, and employment routinely appears as justification for particular economic developments. In Mary's work on waste, for example, small businesses justified their efforts in part because of their self-described *moral* role in generating employment (Lawhon *et al.* 2018a). In Tyler's research, job creation continues to be used to rationalize extractive developments, such as mines, in northern Indigenous communities (McCreary *et al.* 2016).

In both places, it was widely accepted, throughout various industries, that more people than are needed are hired for various roles. In other words, the productive capacity of labour was often seen as of secondary importance: pivotal in ongoing explanations of development is a moral responsibility to provide employment. The centring of work, rather than incomes, livelihoods or even production can even be found explicitly in international discourse, neatly captured in Sustainable Development Goal 8 which aims for "full and productive employment, and decent work, for all women and men by 2030".

Employment, thus, is routinely framed as a goal, not a means to an end!

Where did this moral connotation come from? Why, in a world in which we already produce more than enough of most things, is working – working more, and making more work to do – considered a good thing, an *inherently good* thing?

When viewed in a global, historical context, this approach to work is rather bizarre, profoundly atypical. If the reader will forgive a rather broad generalization here: in most cultures, places and times, when enough work has been done to provide basic material welfare, people stop working, share what they have and spend their time in more pleasant ways.

So we (separately, at this time) went looking for answers. Our searches were rooted in different places, but pointed towards a parallel set of inquiries.

Through colonial encounters, in northern Canada and southern Africa, Christian missionaries sought to instil a belief in the virtue of work. This belief also travelled in the minds and hearts of many secular colonial subjects who sought to "civilize" racialized "others" through demanding their labour. For the most part, in the colonies and the core, there was not explicit and overt collusion between the various churches, states and industries but a long and slow social process in which various parts of the "civilizing mission" came to merge together (Comaroff & Comaroff 1991).

A considerable amount of ink has been spilled over whether religious or economic pressures were more important, a debate often framed around the works of Marx and Weber. Following Weeks (2011), we find no reason to separate these but instead find complementarity and benefit to interpreting these pressures as mutually reinforcing, varying in importance over time and space.

In southern Africa, colonial and, later, apartheid farms and mines needed workers. Taxes were created and land was enclosed in order to create a mobile population without livelihoods (Cooper 1996; Freund 2013; see Chapter 3). This is a familiar telling of history.

These newly created populations of potential workers did not easily submit to the existing options. Men often worked in the farms and mines for short times, retaining some sense of agency and mobility, much to the frustrations of capitalist and colonial employers and states. This part is less familiar in the public telling of history, yet for historians is a well-established pattern. We emphasize it here as a reminder that instilling the modern work ethic was not easy: it was and continues to be resisted.

It continues to be uncomfortable for us (Mary and Tyler) to say "colonial history is replete with resistance to work", for this resistance to work was routinely used as justification for framing colonized peoples as immoral, lazy and needing to be taught otherwise. Calling attention to the resistance makes us nervous because of the ongoing valorization of work by many people in many places.

But for the argument developed in this book, reluctance on the part of many colonized subjects to embrace the modern work ethic needs to be named, and credited as a source of inspiration for our changed ethics.

It is the moral judgements that critique resistance to work that need to be re-evaluated and rejected.

Historically, across both the global north and south, many people (whose politics remain difficult to name, see Chapter 3) living in a changing political economy sought self-sufficiency rather than paid labour. When English peasants were first dispossessed of the land, they sought other means of sustenance via begging, vagabondage and robbery. As Marx describes, their desire for autonomy had to be suppressed, directing the dispossessed "by gallows, stocks and whippings, onto the narrow path to the labour market" (Marx 1993: 507). Concerns with social unrest among the impoverished led to the provision of minimal aid. However, as Piven and Cloward (1971: 22) argue in their classic text, "relief arrangements deal with disorder, not simply by giving aid to the displaced poor, but by granting it on condition that they behave in certain ways, and most important, on condition that they work". Workhouses sought to provide minimum relief, conditional on regimes of degrading discipline and stripping life to basic necessities. While aid programmes have expanded in subsequent years without requirements to work, it has been doctrinaire that these expansions of benefits – to the elderly, caregivers and orphans – conditionally incorporate only those who are unable to work (Ferguson 2015; Slorach 2015). Again, there are widespread parallels here with the enclosure of the commons and the refusal by many to participate in waged labour in historical and contemporary accounts across the global north and south.

Over time, the belief that there was a connection between working, modernity and a good life seemed to be adopted and even experienced by some (but crucially not all) colonized peoples (see Fouksman 2021a, 2021b). We can see evidence of this in many of the narratives of African independence movements: rather than critique the colonial value system, many sought to be incorporated into it, centralizing the modern worker as the pinnacle of liberation. He (and yes, he) would build industries in modern cities, modernize agriculture and create, in short, new and independent modern nations. The embrace of a modern work ethic was also present even for those more sceptical or partial in their adoption of modernity, evident in the valorization of labour in Nyerere's socialist Tanzania (Brownell 2020). The lauding of the worker was not exclusive to politicians or social movements, but was and often continues to be perpetuated by critical scholars writing about such movements and their histories (Barchiesi 2011; Lawhon *et al.* 2018a).

Across the global north and south, there were exceptions and those who pushed back against this reification of work. Paul Lafargue, a mixed-race creole who was born in colonial Cuba and rose to prominence as a French socialist leader (and the son-in-law of Marx), demanded a socialist politics that decentred work in the nineteenth century. In *The Right to Be Lazy*, Lafargue called for a politics of leisure and self-development rather than labour. "A strange delusion possesses the working classes of the nations where capitalist civilization holds its sway", he argued. "This delusion is the love of work, the furious passion for work, pushed even to the exhaustion of the vital force of the individual and his progeny" (1907 [1883]: 23). The cult of work, Lafargue held, propelled people towards overproduction and the creation of fictitious needs. Rather than improving the industrial system, he called instead to break its chorus, to compose "new songs to new music" (1907 [1883]: 47).

A century later, Andre Gorz repeated Lafarge's call in a new register. Challenging Marxist orthodoxies that privileged the labourer as the revolutionary subject, Gorz (1980, 1982) held that radical change would not originate from those who identify with their position as a worker but more general alienation from undesirable work. Instead of the continual focus on the power of organized labour and the struggle to improve labour conditions and increase wages, Gorz pushed for a radical embrace of a UBI as part of building more sustainable futures (Van Trier 2021).

In sum, the modern work ethic is not an inherent human condition. The interwoven beliefs that labour and livelihoods must be linked and that work has moral value beyond its use value are a cultural anomaly, a product of the intertwined powers of capitalism, Christianity, colonialism and modernity. The extent to which the modern work ethic is still widely believed – and repeated as a truism rather than a socially created, politically reinforced and economically interested imposition – is evidence of the success of its advocates and its ongoing embeddedness in contemporary thought.

Why does all this matter?

Identifying this ethic helps us to see why the predictions of the past that work weeks would get shorter and shorter have not come to be. (Most famously, nearly 100 years ago Keynes suggested that we would likely all be working a 15-hour week by now.) Our world is filled with unproductive work, some of which was explored by Graeber in his book *Bullshit Jobs*. Graeber's (2018) descriptions swiftly debunk the presumption that work is necessarily a source of dignity, providing a plethora of examples of people who do not find their own work to be making a contribution – to anything beyond a paycheque.

Such inefficiencies seem to fly in the face of rational capitalist economics. Yes, capitalism induces overproduction, but this does not explain why inefficient busywork continues to be so prevalent, even paid for by capitalists.

Graeber (2018: xvi) suggests, "The ruling class has figured out that a happy and productive population with free time on their hands is a mortal danger". We agree with his core contention that if people had enough *and* more time on their hands, they would demand even more social change. This is a fundamental correlation for the arguments we develop throughout this book. Yet there is, we suggest, more to these relations than elite conspiracy. For the ongoing influence of the modern valorization of work continues to be embodied and internalized not only by elites, but by many ordinary people including workers *and even many radical labour scholars and activists.*

Undoing the modern work ethic is a process that transcends class politics, requiring unlearning the norms it has produced.

As but one example, in their book also focused on a future of enough, Dietz and O'Neill (2013) advocate for a minimum income and also support a jobs guarantee. Why do they believe in the need for this dual approach? Jobs, they argue, are necessary to provide a sense of worth. Ironically, however, this point comes after an extended anecdote of a volunteer working in the sustainable food industry who loves this work, but supplements her volunteering with waged labour to pay her bills. As we will detail further below, there is a bizarre circuitousness to defences of job guarantees: rather than enabling this worker to continue her volunteer work by providing a basic income, the authors insist that her voluntary work should instead, somehow, become paid.

How funding will be generated for what the authors and volunteer agree will never be, on its own terms, monetizable work remains largely unaddressed. The authors broadly assert a need to value this work differently, presumably collectively paying the volunteer to do this work. Yet such arguments are underpinned by deep problems. In the next section, we tease out the values and calculations that underpin paid work, and the limits of proposals to create more paid work through a jobs guarantee.

Why not advocate for more modern jobs through a state jobs guarantee?

A jobs guarantee, like basic income, is an idea that has come and gone in different places and at different points in history. As its name implies, it is based on the idea that the state provides work for any who want it, and that this job provides a living wage (Tcherneva 2020). The proposal of a guaranteed job aligns with contemporary discourses around the desirability of full employment as a solution to poverty and an effort to redress inequality. Some have advocated a jobs guarantee as a core pillar of sustainability transitions (Aronoff *et al.* 2019), including advocates of degrowth (Mastini 2021).

We have two key concerns. The first is that "jobs for all" empowers the state to define what work matters in problematic ways. The second is that the pressure to create jobs for all has perverse impacts for sustainability, actually making it more difficult to reduce environmental impacts.

Let us elaborate.

Contemporary proposals for a jobs guarantee suggest those displaced by technological change need new employment. It is based on the idea that a sustainability transition will create new jobs, but that there will not be enough paid labour without state intervention.

There is much intuitively sensible about this position, and it tightly accords with the modern imaginary of work.

While a jobs guarantee responds to contemporary concerns with sustainability, the model of the state provision of work is not new. Instead, much of the discourse here resonates with decades of (often neoliberal) demands that welfare be replaced with workfare.

In the past three decades, welfare-to-work initiatives have diffused globally. Paradigmatically, these transitions were rationalized on the basis that welfare bred dependency and working was morally preferable. However, actually existing experiments with workfare did not generally empower or actualize workers. Rather, they produced drudgery and exploitation within a system rife with inefficiencies and corruption. Ultimately, workfare functions as a new disciplinary regime to deter welfare claimants and normalize low-paying, precarious jobs (Peck 2001; Mcdowell 2004; see Seekings 2006; Alik-Lagrange & Ravallion 2018 for a comparison of workfare and cash transfers). Thus, our first concern is political – that a jobs guarantee would not empower the working class but rather re-establish relations of political and economic dominance. Some might suggest that a radical socialist response would mean that a more powerful modern state would do workfare better; in keeping with arguments developed throughout this book, we again are sceptical.

We agree that there are and will continue to be new jobs that will arise through the creation of a green economy. The state and citizens should support the emergence of a green economy (see Chapters 3 and 4)! There will also be a need for more research, more planning, new governance, regulations and oversight. Some forms of employment by the state and private sector will surely increase during a transition.

To state very clearly: we have no qualms with people being employed to do necessary and useful work!

However, a jobs guarantee creates a complex set of incentives that will make it difficult to reduce unnecessary, unsustainable activity and prioritize making places for life, not work. Many of the new jobs people are being trained to do are the types of economic activities that a truly radical postcapitalist green economy would eliminate. Investments in retraining, and defending employment, incentivizes the creation and protection of superfluous economic activity. We agree here with degrowth advocates that it is easier to achieve sustainability if certain types of activity are reduced or eliminated; a postcapitalist world does not need millions more computer programmers, and sustainability can be better achieved by phasing out rather than greening certain parts of the economy.

Moreover, as has been shown throughout history, there is misalignment between the skills required for old and new work (Bix 2000). There are already extensive efforts to retrain workers for new types of jobs, efforts that, despite much fanfare and funding, have largely been unsuccessful. Rarely are older workers from displaced industries able to effectively transition to emerging opportunities. This dynamic is exacerbated when new jobs are remunerated at lower rates than previous employment, as displaced workers often consider new opportunities degrading and would rather struggle to achieve self-sufficiency than submit to what they consider substandard employment (see Selingo 2018; Schuppert 2019). Moreover, many of the jobs created through green jobs initiatives will be temporary: as with Chris's employment described in the Introduction, it takes more workers to build things than maintain them, especially if they are built well.

Most importantly, from the perspective of enough for all, there is no need for people to undertake much of the work that retraining projects are preparing workers for.

Further, many of the new jobs produced through a green economy are well suited to automation: they are often difficult, repetitive, drudgerous and dangerous jobs (see Chapter 3). Emergent employment options could be changed, with effort put into creating good jobs. Repetition can be reduced by reworking the factory, better linking individual workers to particular products. Safety measures can always be improved. But our fundamental point of entry here is not whether work can be made into better work: of course it can, and greater social control can open possibilities for democratically deciding to prioritize good work over production efficiency. In some cases, this will make sense.

But such deliberations can only be reasonably had without inducements to create work in order to justify incomes.

What about shortening the work week? If there is less work to be done, it might well seem that the most straightforward strategy is to total it up and divide it equally and base the payment on what is needed for a living wage. In this context, we might all be soon working part-time and earning a living wage from it.

We understand the appeal of evenly splitting work. We are deeply sympathetic to the idea of a modern and rational world where what work is needed and counts is agreed upon collectively. Yet without an incredibly powerful, interventionist state that can appropriately measure, calculate and monitor work, this is simply unlikely to be a viable path. First, there is a muddy set of questions around labour economics and the impacts of a well-paid state guaranteed job at a living wage for a short work week. It would likely induce an increase in wages across the broader economy, resulting in further automation, reduced private sector employment and an even greater burden for job creation by the state. In South Africa, this is a real and explicit concern, and is why, despite demands from the public and unions, state work programmes have generally not been well paid (Seekings 2006). A radical socialist response might be to fully collectivize the labour market, placing all labour fully under collective control, a modernist response subject to the many critiques we raise elsewhere in this chapter and book.

Second, not working for wages – not needing to live near work – also enables more creative and sustainable worlds. It is well established that shorter work weeks reduce ecological impacts (Lewis & Stronge 2021). The reasons for this vary: people spend less time in transit, buildings are heated and cooled for fewer hours, there is reduced pressure for quick foods and their associated waste. Reducing the number of workers would likely increase many of these measured effects, but also have more substantive impact. As but one example, it would radically reshape work-based mobility planning and infrastructure, opening up possibilities for creating sustainable human places centred around life, not waged work and commuting.

These are real but not necessarily insurmountable challenges of a jobs guarantee.

Our third and fourth – and more fundamental – concerns with a jobs guarantee are that it continues to valorize employment as a good in itself and collectivizes decisions over what work matters and whose work counts. Together, these result in demand for a policy that would unnecessarily empower states and reduce the scope for self-defined activities.

The jobs guarantee, like much of labour scholarship and activism, treats "work" as a modern category that is objective, measurable and substitutable. It is not that there cannot be collective answers to the question of what work matters: there have been and could always be collective answers. A collective approach, however, reduces autonomy over how time is spent. This is delicate territory, and our wider argument in this book is largely in solidarity with the politics of those who have worked to change social perceptions of the kinds of work that has value. Yet a jobs guarantee, in even a truly democratic society, must necessarily undertake the impossible task of justly determining which kinds of work "count" and which will not.

Here, we draw on a thread of feminist scholarship that has walked through the difficulties of turning carework into state recognized, metriced and monitored labour (Cudd 2006; Weeks 2011, 2020). Some have advocated for expanding the umbrella of what counts as labour, then measuring and calculating, determining which types of carework should be worthy of a publicly funded living wage. Many, for example, have advocated for a modern approach to carework in which parents are provided with parental leave for a set amount of time, after which children are expected to enter a measured world in which their guardians work a set number of hours while children are cared for by professional careworkers. Others argue for wages for housework (see Federici 1975). For many who have internalized a modern imaginary, these are understood to be good, desirable solutions: they make it easier for women to work, or count the work that women do. For our consideration here, similar arguments would apply to creative work and ecological work, in which people are paid for set hours or outcomes, making it easier to get an income from creative and ecological labour.

Yet what carework, creative work and ecological work would count, and how to count and monitor it, is fraught with ethical ambiguity. So fraught that many have pulled back from these types of solutions (Weeks 2011, 2020)!

Some parents or guardians, for example, prefer to care for their own children and limit or not use professional careworkers. Mary has, for years, worked a rather convoluted schedule to limit the time her kids spend in institutional care or with home-based careworkers; quite bluntly, I have been fortunate to be able to do this, and I am clear my kids are happier this way (which is not to say other children or carers would be!). Before the pandemic, I regularly paid for someone else to do some of the other domestic work. The modern solution of free or subsidized childcare provides no help here, but is an incentive to use institutional care rather than get help with cooking and cleaning. Beyond my home, the Covid-19 pandemic created conditions for many to experience and reflect on the benefits of more time together. We do not wish to romanticize the struggles of parents during the pandemic. Instead, we use this example to illustrate the rethinking it induced, and the ways in which a modest politics provides alternatives. Nor do we wish to critique institutional childcare or careworkers: these options work well for many already, and of course they too can be made better! What we refuse is the universal, the idea that there is a single solution that works for all children and their carers, the notion that the best strategy is for the state to fund a universal solution.

Further, counting the labour of carework, creative work and ecological work is replete with dilemmas. This point has been demonstrated repeatedly by feminist scholarship, for it subjects the home to the determining eye of the state (Weeks 2011, 2020). (If I keep my house tidier, should I get to claim more hours of work? Who will determine if the children's clothes are unwashed?) Similarly, while there are good reasons to support state-funded creative labour, subjecting funding for the arts to the approval of the state has been fraught in practice. And again: how, and under what circumstances, should the kinds of practices necessary for emplaced sustainability – gardening, composting, cleaning and sorting recyclables, cultivating attunements that enable us to notice change – count (see Stokes & Lawhon in press)?

In this context, abstract platitudes of what counts are easy:

Of course carework should count and so should creative work and labour that makes our world more sustainable!

Of course the details can be redressed at the level of democratic participatory citizenship! The modern state can become ever more elaborate and complex in order to build a bureaucracy that can fairly manage such complexity! Or maybe, we can come up with general principles and trust people not to abuse the system.

Yet these sorts of details are precisely the kinds of details that have derailed the modern welfare state.

Citizens and states become preoccupied with ensuring that funds are not wasted, resulting in elaborate rules and bureaucracies to monitor and ensure that public funds are being used in ways deemed morally appropriate through hegemonic norms (Funiciello 1993; Poppendieck 1998; Peck 2001; Mcdowell 2004; May *et al.* 2019; see Chapter 4).

Those of us who have never had our domestic life scrutinized by the state may well struggle to imagine the problem here. For those optimistic about the ability of the state to determine what work ought to count, we urge you to read the literature, to hear the stories of countless people who have felt degraded, exposed and judged by these processes. As we have noted in more detail elsewhere (Lawhon & McCreary 2023), it is no surprise that support for basic income (rather than accounted welfare) is stronger among those already monitored and evaluated by the state. As with the conditions of work noted above, we do not doubt that better practices are possible. We reject the idea that justice should be pursued through such a system.

In short, modern norms work well for those whose lives fit collective norms: full-time work and full-time childcare would help those who embrace the latest version of the modern work ideal. They induce the normalization or punishment of those who do not, often creating unforeseen and unfortunate consequences.

If a job guarantee were the only way to provide livelihoods to all, these limitations on sustainability and autonomy might be justifiable and preferable to the world we have. It would replace some market-based decisions about what work is valuable (which are significantly flawed) with collective ones. This might often be better, but still is likely to reinforce hegemonic norms, valorizing and pushing work towards what is legible and quantifiable.

Yet, as we show throughout this book, rejecting modernity – including its instantiation in both the modern work ethic and a jobs guarantee – does not mean we have to accept radical inequality.

The ethics of work outside modernity

Above, we traced the emergence of the modern work ethic, and suggested that it is not the only way in which to understand the relationship between work and value. While it may well be the most familiar to most readers, it is surely not the only ethic of work. Where might we look to find alternative understandings of a good work ethic?

Our search is again informed by places where the modern imaginary has always been partial. A key intervention in our own reading is Comaroff and Comaroff's effort to disentangle the overburdened category of work. Writing in 1991, they point to the significance of the distinction in the language siTswana between working for the self and working for others: there are actually different words for these two types of work. These differences are not presented neutrally, but instead there is a clear preference for the agency and autonomy of working for the self.

This point is widely shared among libertarian, anarchist, diverse economy and some postdevelopment literatures, as well as many public surveys across the global north and south (Scott 2012; see also Dawson 2021; Lawhon *et al.* 2023b). It is impossible to know how people might act in a world of enough, but it is likely that *some people* might find *some kinds* of work pleasant; *some people* might well enjoy *some kinds* of creative, productive and reproductive work; *some work* might reasonably generate a feeling of pride *for some people*. In short, not all positive associations with all work ought to be attributed to colonial modern capitalist Christianity.

And we say clearly (and loudly, should our present and potential employers read this chapter) that we do not intend to critique all forms of work nor those who enjoy their work. We like, at times even love, our jobs! And Mary writes to her children: I like some parts of carework!

For example, in a recent project on waste livelihoods in South Africa, Anesu Makina spent several months undertaking research with waste pickers in Tshwane. Counter to public perceptions, many waste pickers report very much enjoying their work. They were not those who could not find work, but had often rejected more formal labour arrangements (Makina & Lawhon in press). To be clear, waste labour in South Africa is fraught in many ways and our point here is not to romanticize what are often hard lives. But research participants report that the serendipity and autonomy (albeit entangled autonomy) of waste picking are part of what drives them to do this work, despite often having other (also imperfect) options (Lawhon *et al.* 2023a; see also Millar 2008; Reno 2016; Wu & Zhang 2019).

What all this means for how livelihoods might be created is, however, not entirely clear. Rejecting the modern work ethic does not mean that there is no work to be done (there is), nor is it to deny that some kinds of work might be enjoyed by some people (likely), nor is it to reject all associations between payment and labour (some work ought to be paid). Nor do we find much use in a vision that everyone might directly obtain what we need through self-directed activities; throughout our very collective human history, this has never been the case, nor in the context of environmental degradation and existing population densities is it a particularly plausible answer.

Rather, the central question in a world of enough is: if a just and sustainable world of enough does not need the work of everyone, how is everyone to be assured a livelihood, as well as access to resources and time to engage in emplaced sustainability? In what follows, it will come as no surprise that we propose a universal, unconditional basic income and affordable, collective, modest infrastructure as a modest approach to livelihoods.

Modest livelihoods: universal, unconditional basic income and affordable services

A modest approach to livelihoods is premised on enough for everyone, mindful that enough is not a fixed amount. There will always be fluidity and a need for public dialogue about what is enough. In this context, targeted programmes based on knowable social conditions (whether by the state or another collective) simply do not make sense (see Chapter 4 on universal basic services). Providing unconditional income rather than guiding recipient behaviours through provisioning and conditionality, a UBI enables autonomy (albeit entangled with social and ecological relations) rather than reinforcing social norms.

For many modernists and arcadians, this autonomy is a limit of a UBI: it removes the state and collective ability to induce "good behaviour" and produce good subjects. A modest approach insists that this is a *positive attribute of a UBI: no collective ought to be able to deprive anyone of a basic livelihood because they do not conform to social norms.* This removal is precisely why a UBI accords with a modest view of humanity as indeterminate, heterogeneous and unfolding.

A modest approach means it is not our collective task to determine what types of activities ought to be valued, nor who ought to value them. Instead, a just but entangled autonomy means enabling people to choose, as freely as they can, how to live their lives, contributing to a world in which people have more choice over how they spend their time. These choices can be made more freely on two important fronts: the first is by rejecting social norms that valorize work as an end in itself. The second is with the assurance of a basic livelihood not dependent on labour.

A modest state ensures that no one has too little or too much in a context in which environmental constraints preclude abundance (see Chapter 4). A modest approach means creating an economy in which people can work, own and participate in collectives without their livelihoods depending on this (see Chapter 3). It recognizes there is not a clear line between what work should and should not be undertaken. There is and always will be some work that needs doing, and at times people will come together to decide that certain work needs to be done, sometimes through and sometimes outside the state. They will also decide how to incentivize and/or compensate those who do it. But there are also many activities, ranging from gardening to painting to playing cards that are enjoyable and desirable activities for many people. At times, these may be useful, productive and reproductive to some, and for some, and at times, not. A modest approach rejects efforts to create a metric for adjudication and counting of labour, determining what counts as enough to deserve a livelihood.

In making this argument, we are indebted to feminist insistence that carework be recognized as work. We are also inspired by those who insist that it cannot be adequately valued by the market nor can it justly be subjected to quantification and comparison, and that there are deep problems associated with the monitoring required to make such a system feasible. These arguments for human carework can also be expanded into much the work of emplaced sustainability, work that also cannot be objectively measured and accounted for. In keeping with the wider argument throughout this book, therefore, we do not seek to expand the reach of modernity to capture more people and more work. In this sense, we adopt and expand on Weeks's (2011, 2020) reconsideration of the problem of work and support her conclusions: a universal, unconditional basic income is the best way to recognize and value the diversity of work *without* collectively counting and making normative valuations of it, enabling people to live outside a world where their lives are governed by metrics.

In this context, a modest approach to livelihoods is about creating a context in which everyone has enough, but the contents of the good life are not collectively determined. A redistributive UBI embodies a modest approach to livelihoods in two key ways: by removing conditionality, it does not increase the biopolitical power of the state; by providing cash, it enables people to determine how to build their livelihoods (Lawhon & McCreary 2023).

This does not mean a UBI is without flaws. A UBI does produce the dependency often critiqued by scholars who seek change outside of the state. Pragmatically, we think this is a risk worth taking in the imperfect world we have. For most people, economic redistribution through the state would increase their everyday experience of autonomy (albeit an entangled one) and the benefits of using the state for economic redistribution cannot be matched through any other means. It would, of course, be permissible for an individual to reject a basic income and live without this dependent relation on the state. More substantively, however, the very premise of embedded autonomy is that relationality is a necessity; we are all, always, necessarily dependent on our relations to some things. Living in an entangled world means some kinds of dependency (Ferguson 2015), whether it is on a forest to provide fruit, a farm to provide wheat, our family to provide labour or upstream neighbours to not wash waste downstream. In this context, we consider the dependencies associated with a UBI to have less risk of abuse than the entanglements created through modern statecraft or being subject to the collective will for the direct provisioning of basic needs (Lawhon & McCreary 2023).

A UBI is no magic bullet. If nothing else in the world was to change, we agree that a UBI would not create human flourishing and autonomy. Yet our fears about dependence on the state are mitigated by contextualizing a UBI within the wider set of changes we have outlined throughout this book. We are convinced that a UBI has the potential to be part of upending existing unsustainable and unjust relations in a world demanding change.

Conclusion

Conventional labour scholarship and politics have often been underpinned by a modern imaginary, critiquing work that is not modern and pursuing increased access to modern work. While scholars have routinely criticized the modern imaginary of the economy and state, the full implications of the modern work imaginary has only begun to be examined within the academy. This powerful imaginary has and continues to be believed by many across the political spectrum, often unstated and serving as a foundation for policies, activism and moral judgement. It is rooted in a twinned argument that links labour and livelihoods in the valorization of work.

The modern work imaginary has, however, not been without its critics. In practice, it has always been resisted. A small but significant body of work, often rooted in postcolonial and feminist theory, has pushed to articulate and critique the normative assumptions that underpin analyses of labour and the political strategies that emerge from this modern ideal of work. Critical scholarship that has shown the social construction of the modern work imaginary has had limited impact on politics as yet: it has failed to significantly disrupt, let alone displace, the hegemony of the modern work ideal. Instead, labour politics often embraces and builds policy proposals that implicitly reinforce this modernist imaginary. This may be, in part, as we have argued throughout this book, because of the wider hegemonic discursive power of modernity and the paucity of alternatives.

In this chapter and more broadly throughout *Enough!*, we argue for displacing modernity and modern work, seeking to identify what a modest alternative might entail. Through our intervention, we aim to open political possibilities for imagining other relations and offering another story about labour and incomes. A modest approach starts with the tautology that work is useful when work is useful, and also recognizes that some people find additional value in some work. Some work will always be necessary, and collectives (whether the state or not) should sometimes decide there is work that needs doing, and determine how it ought to be allocated and compensated. Yet such decisions can be most justly made when basic livelihoods and labour are disentangled.

We propose that modest livelihoods can most straightforwardly be achieved through the assurance of affordable basic needs and a universal basic income. A UBI better accords with the modest approach we have developed in our wider argument than competing proposals for a jobs guarantee or universal basic services (see Chapter 4). Were a jobs guarantee or UBS the only way forward, it is preferable to the status quo in which markets undervalue many kinds of labour and overvalue others.

But there is another way, one that is politically, pragmatically and ethically preferable.

A UBI enables people to spend time on carework, creative work and the work of emplaced sustainability that the capitalist labour market poorly compensates, without collectively deciding, counting or monitoring it. A UBI means that people can choose to be artists, undertake carework, and support their communities according to their own values. It means that there is time and security, enabling people to choose the work they want, including the work of building worlds we might want.

6
ONWARDS

Through *Enough!*, we (Mary and Tyler) have sketched emergent possibilities for a modest politics of sufficiency in an uncertain, indeterminate world. We began the book with the argument that, in a world of scarcity and surfeit, there can sustainably and justly be enough for all. Drawing on social and ecological literatures on uncertainty, we differ from modern approaches that treat "enough" as a universal or mathematical determination and insist that "enough for all" cannot be universally known or collectively, directly allocated. Instead, we develop a modest approach to enough for all, treating it as a political ecological ambition that can motivate and guide a modest approach to the state, economies and livelihoods. Further, a sustainable version of enough for all can only be enabled through emplaced attunements with the ongoing changes in our surroundings. This book is our effort to tease out the implications of these points for a modest politics, an approach that subtends trenchant political fault lines, opening opportunities to forge new alliances and chart pathways forward.

The arguments we develop here did not begin with a grand theoretical quest. We are trained as deep case study political ecologists who linked and critiqued the imbalances of power and unsustainable practices we saw, reading, listening and learning as we engaged with post- and decolonial theory, livelihoods and infrastructures in North America and Sub-Saharan Africa. In response to ongoing questions from research participants and students, as well as our own politics, we increasingly felt the need to think beyond – while still dwelling in – our cases to answer the question "what are we for?" (cf. Ferguson 2010).

We failed to find ready-made answers we believed in. Theorizing in dialogue with scholars and activists in the global south, Indigenous communities within the global north, as well as our students whose politics criss-crossed existing spectra, enabled us to – demanded that we – grapple with the limits of dominant approaches to redress injustice and unsustainability. It meant we needed not only to unlearn foundational core assumptions, but to rethink our values alongside our politics. As we searched, we increasingly found ourselves part of a wider movement within and beyond political ecology looking for ways to theorize and imagine the future otherwise.

Writing in dialogue with, but not only to and for such people and places, *Enough!* asks: what does justice look like when we claim there is enough for all but that the collective (neither the state nor the commons) cannot accurately know and fairly manage the needs of each person? What does sustainability mean amid uncertain knowledge of, and relations to, the earth and its constrained but fluid capacity to support us?

This book became a grander vision than we anticipated: through it, we articulate our aspirations to nothing less than a reimagining of political possibilities, desires and demands. Doing so, our intent has been notably immodest in the vernacular sense of the word, aiming to foster the development of a new radical politics that differs from orthodox articulations.

Yet as we noted in our introduction, the more we have discussed our arguments in and beyond the academy, the more our sense is that much of *Enough!* feels intuitively almost familiar, resonating with the insights and experiences, hopes and dreams of many of our students, colleagues and post-pandemic public conversations. What we propose here is radically different from the world we have and modern and arcadian visions and vocabularies. But our hope – rooted in many conversations – is that much of what we are doing is providing language, analytical tools and a broad political narrative that resonates with already-existing intuitions and aspirations.

We are clear that the pathway we propose here is not a swift and easy one. We understand the need for urgency: we too are tired and want to see change! Modest ideas and experiments are already out there, and can be learned from and built on, but widespread change will take time. Were a modernist pathway of a strong state and the rolling out of big technological change able to be successful, this would be a quicker avenue. But, as we hope to have shown in the previous chapters, this is unlikely to be a just and sustainable pathway. Some might be able to build walls and provide temporary protection for the elite, but ultimately modernist efforts to control are short-sighted, deeply flawed, based on a fundamentally inaccurate understanding of ecology and each other. They have induced ecological crises and unruly social responses, revolting against such order. Paraphrasing Mann and Wainwright (2018), neither Eco-Leviathan nor Eco-Mao will ultimately be sustainable, and certainly neither are just.

To ignore modernity's flaws in order to hope for rapid change – to ultimately opt for a swifter path that will not work – is not a better option.

A modest approach is incremental, not immediate. Ecologically, it will not ward off the worst of what is to come in the short term. We defend the wider arc of argumentation here because the worlds that can be built through a modest approach leave most people better equipped to live well in uncertain ecologies, more able to grapple with changing climates and relations, than arcadian or modern approaches. A modest approach will not easily and quickly fix everything, but there are no other options we (Mary and Tyler) have found convincing.

In the remainder of this chapter, we first tease out explicitly the fields and politics which we see as most aligned with the wider project we have outlined, and potential points of friction in creating a world of enough. We then outline the broad contours of what we imagine a modest world of enough might entail, including the kinds of political ecological subjectivities that might emerge and be cultivated. We conclude with a brief reiteration of our main arguments and hopes for the future.

Allies, alignments and frictions

Returning to the question of "we" from our introduction, who do we anticipate might find the arguments we outline here compelling? While our naming of a modest imaginary and a politics of enough is novel, it builds on a long lineage of precursors and, as we hope to have shown throughout *Enough!*, is forged through reading and speaking in relation with a range of cognate thinkers. Here, we name a few bodies of scholarship (ones referenced regularly in previous chapters) as a reminder of our debts, synergies and allies. For the power we hope to cultivate through *Enough!* is not in the abstract articulation of a unique imaginary and associated politics, but the affinities it shares across an eclectic but not incongruent community.

First, let us be clear that there are people who benefit from the world as it exists, and some profess a nostalgia to return to a modern world of racial, colonial and gendered hierarchies. It is our wager here, however, that many more are jaded by the failed promises of modernity and searching for more just and sustainable ways of being – even if not always with this language. In the global north, the relative security of the Keynesian economy for (typically white and male) workers has unravelled. In the global south, the hope for a radically improved future that accompanied decolonization has waned into postcolonial cynicism. Xenophobic and revanchist politics are on the rise, blaming other groups for contemporary conditions and intensifying a politics of fear and resentment. There is a visceral current of anger underpinning much of contemporary politics. Simultaneously, huge numbers of people, who never embodied a valued economic identity, live in economically alienated conditions, internalizing disregard into an expanding disbelief in the possibility for improvement. These disinterpellated subjects are not captured by dominant political economic subjectivities but disidentify with them.

It is too much to hope that a book can change all this. Yet the wider political agenda of *Enough!*, we think, might provide some guidance towards a shared framework and more capacious alliances. The political project we outline here does not directly attack all of the kyriarchy and revanchism that can be found in our world but instead charts a pathway forward in which many might see themselves. This is not to justify the kyriarchal or revanchist responses, but to subtend some conflicts in order to forge a pathway forward that may attract support across old political fault lines.

Let us be clear: many privileges will be lost in the pursuit of a modest world of enough. Our wager is that, for most people across demographics, identities and old political spectra, newfound and deepened freedoms and security outweigh these losses. The widespread disenchantment with the world we have provides an opportunity to build a collective platform across many seemingly disparate axes, drawing on a foundational shared thread across wide-ranging places and peoples that demand a world in which we all may choose how to live.

Those who continue to firmly believe in modernity might find little to like in our text. That said, for the many, many modernists whose faith is wavering – for those who see the widespread ecological crises and abuses of modern states but struggle to let go of hope for a better world – we hope to offer a vocabulary and vision that makes it easier to let go of a flawed imaginary. We have much sympathy for modernist thinking and those who continue to be attracted to the possibilities it offers, for those who see the cracks in modernity but hope to patch them rather than build a different narrative. We are sympathetic to the urge to retain a hopeful term and fill it with new content. Ultimately, however, *Enough!* is not another text focused on what is wrong – many such texts have been compellingly and convincingly written many times. Our strategy is to speak to those already sceptical, who hold on because they lack a compelling alternative. *Enough!* is meant to help sharpen our critique and, most crucially, cultivate a vocabulary, vision and community of people willing to consider what comes next.

We hope this makes it easier to leave old debates behind.

In this context, our politics do not look much like modernist visions of socialism, the forms of state-planned socialism that so many think of when they hear the term. Yet many who use the term today use it in a more capacious fashion, as a label for what Wright (2019) suggests is an economy subsumed to social demands. We do not use the term "socialist" to describe our vision and, particularly given its ambiguity and ongoing fraught weight with our students and in public discourse, we disagree with Wright about the political and analytical utility of the label. But we do believe that much of the content of what we propose is aligned with the concerns, visions and politics of many who self-identify as socialist.

Similarly, our arguments do not look much like modernist visions of capitalism, either in canonical writings or the practices that are labelled capitalist in the world we have. Yet much of what we write is in keeping with the aspirations for freedom sought by many foundational pre- and early capitalist thinkers who sought to overcome the hierarchies and problematic norms of feudal Europe, and those who defended capitalism when totalitarian socialism appeared the only other alternative. While we have drawn primarily on other literatures throughout *Enough!*, what we write is deeply shaped by our time spent in politically conservative places and thinking with – and taking seriously – concerns from those who do not align with critical and radical politics. We, too, do not wish for a world in which individuals are necessarily subsumed by a collective. We are also clear that the freedoms sought by those who overthrew feudalism have been widely suppressed by the capitalism-we-have in the world today. We therefore seek political economic relations that accept the impossibility (and undesirability) of transcending all entanglements while also valuing autonomy.

In short, *Enough!* expands and builds on, rather than rejects, the concern with freedom expressed by many advocates of capitalism. We too seek greater autonomy, but not only in the marketplace. We develop a notion of autonomy that is more expansive than that in capitalist political economic thought, and root our politics in acceptance of the inevitability of some social, ecological and economic entanglement. Throughout this book we have thus not critiqued the possibility for progress and freedom that sit at the heart of capitalist ideas, but instead centred the ways in which modernity has limited freedoms, vast economic inequalities constrain choices for many people, and neoliberalism obfuscates who we are (and how we come to be).

While we think the argument developed here has wide resonance, we are also clear about where our closest already-existing kindred thinkers lie. Our ideas are rooted in post- and decolonial scholarship that has long pointed to the limits of modernist conceptions of development. Postdevelopment writers have rejected both the social and material worlds imagined through modernity, and we agree with their concerns. We disagree with the arcadian aspects of *some* of this work, including the presumption of harmonious communities and relations with nature. We also challenge arcadian scepticism of the state, technology and broad notions of progress. We are not the first to make such critiques.

Yet postdevelopment thinking is neither uniform nor static, and we find much inspiration in writers that have reflected on these criticisms and reworked postdevelopment thought away from arcadian visions towards what, in many ways, accords with the wider arguments we develop in *Enough!* Part of our contribution here is to enable sharper distinctions within the wider remit of postdevelopment and cognate thinking in order to differentiate arcadian threads from more politicized, progressive ones, enabling clearer explication of the politics of development. Crucially, we do not believe that this is incongruent with what many thinking around postdevelopment today are pursuing.

Similarly, our arguments resonate with much that has been advocated for by heterodox Marxian thinkers supportive of diverse, embedded economies. Our politics are less focused on the local scale than early versions of this work, but accord with many scholars of diverse economies that have responded to concerns with parochialism. We also attend explicitly to the norms often implicit in this literature, considering parameters through which diverse economies might accord with notions of justice outlined in Chapter 2 (for all diverse and embedded economies are not necessarily just!). Finally, we work to make politics through the state more amenable to some of the concerns voiced across postdevelopment and diverse economies literature by distinguishing between modern and modest statecraft. Drawing on this distinction, we make the case for a UBI that might underwrite economic diversity and equality, cautiously hopeful that the justification we have provided – that it uses the state without reinforcing its biopolitical power and that funds can be used to support non-capitalist economies – might well attract support for a UBI from scholars writing about development and diverse economies.

Like many, we note some shared concerns across postdevelopment, poststructural and neoliberal critiques of modernity. As difficult as it may be to say so today, we believe that political alliances may be forged through this shared concern, and a shared desire for autonomy, including the autonomy to choose entanglements. A modest economy has no space for the kinds of firms that dominate many iterations of capitalism, yet is supportive of the kinds of private enterprise that many interested in freedom and autonomy care deeply about – small entrepreneurs, family farms – and the use of markets to distribute what cannot be justly allocated by collectives (either states or commons). The kinds of practices that accord with a modest approach, then, look in many ways like the kinds of practices advocated by some ecosocialists, and some who use the language of degrowth, and even some aspects of capitalist, libertarian and anarchist thought, even if we do not find these terms adequate for enabling the distinctions that lie at the heart of what we propose.

In this context, our hopes in writing *Enough!* are not so much to call for more action on behalf of justice and sustainability, although we are clear that more action is needed. Nor is it to call for a new social movement to induce a modest future. Instead, we seek to provide some language, analysis and guidance that we hope resonates widely and might shape ongoing political activity. Such reorientation, we believe, enables synergies and wider alliances. This is not to make it sound easy nor to naturalize the process: undoubtedly work will be needed to build interest, coordination, and make modest demands, and what this looks like will vary across different places. The novelty of what we propose here is not so much in these types of details – political education, protest, networking, learning from elsewhere and prefigurative action will continue to be essential tools for those seeking to build a just and sustainable world. We hope to work with many others to undertake the difficult and time-consuming tasks of creating change!

Our argument here is, in short, for a modest politics and set of demands that, we hope, will make it easier – not easy, but easier – to attract people into the process.

Modest world-making and modest political ecological subjectivities (or: what about Chris and Sipho?)

While our focus throughout *Enough!* has been on wider political ecological questions, including what a modest economy, state and livelihoods might entail, the implications of our inquiry spill over into questions of culture, identity and subjectivity. Neoliberalism treats subjectivity as given: it takes for granted that people are a certain way, and works to build a political economic system accordingly. Yet as we have noted in more detail above (see Chapters 3 and 5), neoliberal subjectivities are not given but constructed! Those operating outside of capitalism – and even those shaped by it – do not mimic the ideal rational actor. Instead, as Gibson-Graham emphasize in their explorations of diverse economies, participation in non-capitalist economies creates new entanglements and provides ways to undercut kyriarchal relations. This not only enables greater choice in the market: it enables us to think more capaciously about who we are, what we want, and how we interact with our worlds.

A modest politics of enough, then, is not only about asking people to accept a different quantity of goods, but to fight for a distinct form of political life, one with new and potentially profoundly changed socioeconomic subjectivities. We cannot know what kinds of personhood might emerge if people were assured basic economic security, more time and autonomy, able to worry less about ongoing and impending ecological crises. There are, however, pointers as to how security in our entanglements might shape who we are, what we want and how we relate to each other. In addition to the extensive work in diverse economies building on Gibson-Graham's foundational ideas, research on cash transfer experiments is cautiously suggestive of new subjects that a modest politics may call into being. As we reviewed in our Interlude, the economic security provided by even limited cash transfers shapes – and has the potential to transform – much more than material needs: it impacts many recipients' sense of belonging, shifting entanglements away from particular productive relations and towards a wider social ethic of care (Lawhon & McCreary 2023).

What kinds of worlds would people build if they were less hindered by kyriarchal relations, more secure economically and less fearful of ecological crises?

The political project we outline in *Enough!* is not deterministic, but invested in a hope-filled answer to this question. Delinking competition and fear from basic material provisioning is likely to profoundly reorient the way we relate to each other and the kinds of material lives and entanglements we choose. In the kyriarchal world we have, it is difficult – at times impossible – to disentangle ourselves from pressures that cultivate competitive, fearful, always unsatisfied subjectivities. In a modest world of enough, some may still pursue activities that induce fear and participate in competitive activities: many may still want to climb mountains and win football matches. Our intention is not to judge, predict nor determine what these activities ought to be.

While we believe that the political project we outline here will cultivate different subjectivities, a modest politics is not about strategically shaping the content of personhood. In this, a modest politics is distinct from modern and arcadian notions that already define good subjects. Instead, we insist that it is not ours (neither Mary's nor Tyler's nor the collective's) to decide what the good life entails for others, nor to make security dependent on fulfilling our definition of this. Thus, *Enough!* is not in an effort to *create* particular subjectivities, but to imagine what might be if existing pressures were loosened, and people were more free to pursue activities and entanglements within the context of sufficiency.

In this sense, while the scope of what we propose is grand, it remains fundamentally rooted in a modest take on what can be known about each other and our futures. It is rooted in an effort to provide a political, economic and ecological framework in which many worlds, many aspirations, many visions of what the good life entails can be pursued – albeit never unbounded and disentangled.

In this grandness, however, maybe the arguments we have made feel, at times, a little abstract. They have to us (Mary and Tyler) as well.

So let us, in this penultimate section, return to Chris and Sipho, our composite characters from the Introduction, to think through how their lives and livelihoods might be shaped by a modest politics of enough.

Sipho, you might remember, is a waste picker. In the long future, it is possible that there will be no such role, but let us start long before this moment. Maybe he and Nomvula, the mother of his children, sign a petition to support the expansion of South Africa's already-existing cash transfers system to include Sipho, a middle-aged, able-bodied man without contracted waged employment. Maybe they join the movement for basic income, toyi-toying with comrades in protests and explaining to their neighbours why they support basic income. Or maybe not.

Let us imagine the movement succeeds. What does this money do for them? Maybe Sipho sends most of it to Nomvula (you may remember, she and the children were living outside the city given the higher urban costs of living), holding on to just a bit to make more phone calls to them. Or maybe his family moves to the city, where they can finally live together. Sipho delights in seeing his children each day, sometimes even getting them ready for school. They miss some parts of rural life, but can afford to take the minibus taxis to visit their grandparents every few months. Now that Nomvula lives with him, she tells Sipho she wants to get married – this no longer risks the children not qualifying as recipients of cash transfers – and that he must stop seeing other women. When she relied on Sipho's money from waste picking, she felt it would be imprudent to make such a demand. But now, if Sipho says no, she will still be able to take care of their children.

Maybe Sipho says yes; maybe no. It is not ours to say.

Waste continues to be discarded, and Sipho continues to collect it. There is a notable change in the waste in his neighbourhood: with increased incomes comes more disposable packaging and products, at first. It is less often on the street, more often in the bins (which get emptied more reliably now that people can pay for this service): people have more time now, seem to hustle a little less. He also sees more renovations to homes, more cultivation of vegetables, more sociability on the street. Perhaps surprisingly, a few shebeens close – or at least, this term no longer seems to apply to the multi-generational places that sell healthier, higher-quality meals, and have more space for pool tables and foosball. Many people still like to drink, but the drinking itself changes, feels less like a search for escape and more like the old social lubricant that made laughing come more easily. Although these days, laughing, too, seems to just come more easily, even without a lager.

Maybe Sipho decides he wants to take a few classes and return to school. Maybe university, or training in electrics. But maybe not: maybe the autonomy and serendipity – and now a bit less conflict, a bit more sociability – of his work means he continues picking waste.

Sipho, maybe, still goes to the wealthier suburbs to collect materials: inequality still exists and this is, for now, still where the greatest finds will be. He is usually in place early, but on days he gets the children ready for school, he may be the last to arrive. Fights over materials still happen, territories are claimed and contested, but Sipho worries less and somehow everyone seems to know there is just a little less at stake now. It is no longer a life-or-death search, eating today no longer relies on what is found. He and his comrades take a few more breaks, and share a few more stories along the way. Some of them have formed a collective that agrees to share the findings of the day. Maybe Sipho joins it, maybe he does not.

As he ages, Sipho slows his work down a bit: the children are now grown and, maybe, Sipho now often gets his grandchildren ready for school. He collects a bit more material from closer to home, although he still regularly goes back to the old suburbs where he long worked. There are fewer great finds: even those in fancy houses discard a bit less these days, and sort their materials better. Old relationships still exist: a gardener still saves items for Sipho, young people he has trained over the years continue to work in the area and taking breaks together continues to be one of the highlights of his week.

Maybe Sipho gardens a bit more. Maybe he gets involved in local politics. Probably, he is a bit more relaxed, a bit more patient, works a bit less and cares a bit more. But maybe not.

Maybe this is Sipho's future. Maybe it is not. It is not ours to say.

***

Chris, if we recall, was an environmental monitor for a pipeline construction project in the Canadian north. While it provided him with a relatively lucrative income, it always had a temporal horizon: his employment depended on the continuous expansion of extractivism because more workers are employed in pipeline construction than day-to-day operation. Of course, the climate crisis demanded an end to such growth.

The provision of a UBI provides Chris and his northern Indigenous community with a degree of economic stability, reducing their reliance on extractive infrastructure to provide income. Leaders of his impoverished community previously had signed impact benefit agreements to support pipeline proposals in exchange for jobs and desperately needed funding for community services. However, basic income alleviated the pressure to sign off on environmentally destructive projects. It created the conditions to embolden leadership to fight pipeline proposals and build a more sustainable future. Maybe Chris directly opposed new pipelines. Maybe he did not. It is not ours to say.

The subtending of pipeline politics also helped heal rifts in the community. For years, struggles over pipelines had riven Indigenous communities in northern Canada. The intensifying industrial development had not only provided jobs, it had ecologically eroded Indigenous ancestral lands. This had led to escalating conflicts between community members concerned with unemployment and those fearing the destruction of their lands. With these tensions diminished, there were opportunities for healing. Chris had stopped talking to his cousin Rob after a fight about pipelines. However, with eased tensions, maybe they rebuild their relationship. Or maybe not. The old fault lines, at least, are fading, not passed on to the next generation.

Maybe Chris decides to use the skills he developed as an environmental monitor for ecological restoration work. After decades of extractive development, the ancestral lands of his people were suffering. However, with a bit more ease in life, he invests his time in rejuvenating the fisheries, helping nurture the biodiversity of the land and building emplaced sustainability. The land will never return to what it once was: words like "rewilding" never made much sense in a place that has long been peopled. But maybe, with insights drawn across the knowledge of his ancestors, his university degree and his experience, Chris participates in the slow work of regenesis, cultivating rambunctious gardens of old and new relations.

Universal income support also improved family relations and engendered greater equality in the community. Previously, development benefits had been unevenly distributed, concentrated in the hands of those signing agreements and those taking jobs – typically men. With obligations to care for children, women often lacked these opportunities. Large numbers of women in the community relied on state-granted benefits for single mothers. Unemployed men who did not work in industry generally lacked legal avenues for income and had often sought informal partnerships with women collecting child benefits. This contributed to unhealthy relationships and issues with domestic violence. Basic income helped alleviate the pressure to form relationships out of need. Love increasingly became the basis for relationships.

Maybe Chris proposes to his childhood amour, Lisa, who already has children. Together they could provide a stable environment to support the growth of their family.

Maybe Lisa gets more involved in cultural revitalization work. In the twentieth century, assimilationist residential schools disrupted the intergenerational transmission of knowledge and stopped children from speaking their Native language. Before, some work was already happening, but was hampered by the lack of time and money, and the personal struggles of many community members. Maybe Chris and Lisa home-school their children, allowing them to spend time with elders in the community and to learn their Indigenous language.

Maybe this is Chris's future. Maybe not. A modest politics expands the possibilities for futures such as these, and leaves the details for others to decide.

Reiterating our argument

Recapping our argument, we began by locating our intervention within broader discussions of radical change. We collectively live in a world of widespread inequality and environmental crises, a world in which many are anxious about a highly uncertain future. We (Mary and Tyler) share a disenchantment both with the status quo and with prevalent answers to the question of what to do. Both of us began our engagements with the academy through research with people living in peripheralized places. We struggled with the ethics of not addressing the pressing question of what to do, as well as with the available answers to this question. Our introduction places ourselves with a collective, shared struggle.

Chapter 1 frames our discussion theoretically, charting how modernist aspirations and critiques of modernity shape competing visions for change. We recognize, and build from, the extensive literature on the myriad ongoing social conflicts and ecological crises afflicting our world. Our purpose in this chapter is to review dominant approaches to environmental politics, including their different diagnosis of root causes of injustice and unsustainability and competing visions of a just and sustainable future. To orient readers, we highlight the centrality of modernity to ongoing debates over how to best envision and enact change. Specifically, we chart two distinctive visions of a just green future – ecosocialist modernity and arcadian environmentalism – and review critiques of both. We position the argument of *Enough!* in agreement with the many authors pointing to the insufficiency of both frames, and the need to articulate a vision and a politics beyond this binary.

Chapter 2 provides our conceptual framework and chosen vocabulary. We suggest that a modest imaginary accords with many contemporary approaches to social and ecological theory, including some notable works by political ecologists emphasizing the ongoingness of uncertainty and the inability to justly define the good life for others. We point to common threads across disparate literatures and the analytical utility of differentiating alternatives to modernity and its inverse. In our conceptualization, modesty is not a humble, arcadian, anti-modern politics, but instead is underpinned by a political commitment to enough for all in the face of radical uncertainty.

We also explain how a modest imaginary might provide a platform for advancing a new politics of emplaced sustainability, anticipatory guiding and entangled autonomy. Emplaced sustainability is rooted in an acceptance that what sustainability means is plural, shaped by where it happens. Learning across places is still important, particularly in the context of uncertainty and change. Yet, specific strategies and actions need to be formed in dialogue with those most attuned to particular socioecologies, for the ongoingness of change is impossible to supersede through calculative prediction. Anticipatory guiding refers to a mode of shaping an uncertain future, reflexively learning and adapting to build a better world without the conceit that we are capable of deterministically knowing, managing or controlling everything. Finally, entangled autonomy highlights the need for interplay between the always-encumbered aspiration for freedom and a social and environmental world in which we are linked to human and more-than-human others.

The three chapters that follow seek to translate a modest imaginary into a modest politics through considerations of the economy, the state and livelihoods. Chapter 3 first considers how modernist and arcadian thinking has informed the history of different approaches to the economy. We also trace another line, often forgotten, of enacting aspirations for an alternative economy beyond capitalism and socialism, one that has long existed and continues to trouble ongoing political economic vocabularies and analyses. With this context in place, we then advance the idea of a modest economy, and how it could differently orient thinking around questions of technology, ownership, growth and distribution. Neither advocating for nor opposing growth, a modest economy begins from the recognition that enough is not a fixed quantity but the product of balancing social and environmental concerns in decision-making around questions of technology, labour and distribution.

Instead of the hubris of universal technological solutions or anxious refusals of technology, we argued for a cautious embrace of technology as fundamental to a world of enough. Modest technologies are distinct from modern ones, purpose-built and subject to anticipatory guiding through ongoing modification, and part of attuned, emplaced sustainability. Where possible, technologies that displace drudgerous human labour ought to be embraced rather than feared. Further, we stress the need for a mixed economy, dominated neither by private regimes nor state ownership. We do see reason to phase out shareholder-based enterprises because of their disembedded profit-seeking practices. Beyond this, however, we see scope for a plethora of ownership structures including and beyond private and state ownership, such as neighbourhood associations, consumer cooperatives and worker collectives. Finally, we emphasize that achieving a just and sustainable economy will require redistribution, and an ongoing politics that centres distribution.

How might such economies be created? We turn to this question in Chapter 4, where we emphasize the significance of different modes of statecraft. We again position our arguments against both the hubris of modernist socialism and the rejection of the state as a key actor in the pursuit of justice and sustainability. We agree with longstanding critiques of the exclusions that accompany modernist state projects and oppose the inherent violence that underpins universalizing and homogenizing state practices. Yet we are concerned that eschewing the state limits the horizons for change. To resolve this tension, we consider the potential of modest statecraft, and the kinds of demands that can be made of the state that do not enhance its biopolitical power. Drawing upon our research in critical infrastructure studies, we show how citizens have mobilized to pressure the state to act modestly. Reframing debates over the state this way shifts our focus away from quantitative questions of more and less state involvement, and towards a qualitative consideration of the relationships that statecraft creates between the government and the people. We see basic income as emblematic of the kinds of actions that a modest state may undertake.

Chapter 5 advances a consideration of modest livelihoods and economic security in a post-productivist world. We began with a history and critique of the valorization of work, noting that the modern work ethic rationalizes drudgery. Further, it makes it politically difficult to reduce production in some parts of the economy. Drawing on scholarship rooted in the global south, we argue that cash transfers offer the foundations of new political economic arrangements. Like a growing number of others, we argue that the state should provide people with economic security; drawing on our argument in Chapter 4, we insist this ought to be done modestly. In this chapter, we explain how a modest approach differs from modern efforts to provide directly or assure employment. We conclude with a consideration of how cash transfers exemplify this approach and might enable the modest politics and economy we describe in previous chapters. That said: we are clear a UBI in itself is not a panacea. Instead, we suggest it may play a crucial role as a catalyst and part of a broader modest politics, for it enables time and resources to underwrite the cost of experiments with alternative social and economic arrangements and engage in democratic practice.

In sum, *Enough!* draws on emergent optimism and a sense of possibility in the present. A modest politics of hope runs counter to the proliferation of increasingly fearful politics, and encourages an embrace of the possibilities that inhere to the present. Unlearning can be disorienting, but is necessary for opening our imaginations to the possibilities of the future. We invite you the reader to modestly aspire with us, to aspire to a future of material sufficiency in which people can choose how to flourish, in which we deepen our attunement to entanglements as a means through which to pursue sustainability, and in which political participation is enabled through the redistribution of resources and the freeing of time.

A modestly hopeful conclusion

For a very long time, we (Mary and Tyler) shared an anxiety about the future, a sense of foreboding captured in the litany of longstanding and emergent crises with which we began this book. Our worries were compounded by our inability to articulate a politics we could believe in as simultaneously radical and plausible, just and sustainable.

Thinking through a modest politics and imagining new possibilities has provided us with a new sense of hope. We did not set out to pursue a hope-filled project, yet the more the pieces began – over the course of multiple years – to come together, the more joy we found in teasing out lines of thought, finding congruence with a wide range of ideas, even in typing, organizing and re(re-re-re)writing our thoughts. As we began to circulate these ideas in conversations, conferences and classrooms, we not only found a personal release from our anxiety, but an emergent, multi-generational community of collaborators excited to think beyond the modern and its inverse. We found receptivity among people who desired to engage politics differently, who sought to move beyond hubris or fear and engage in a new politics.

It is in this congruence of thought and feeling, and an emergent sense of community, that we find the most hope for the future. There is a joy-filled sense of possibility that comes from having an idea of where to go, even if we lack a map. It is both with this sense of possibility and an invitation to work together to improve our politics and analysis that we conclude our book.

Modest politics do not represent a complete platform but a call for mobilization around a set of possibilities. Through making modest demands of the state, we seek to make politics anew. Rather than muting or reifying difference, a modest politics makes space for heterogeneous identities and behaviours, including the non-normative, within a shared political agenda. As a demand, it calls into question taken-for-granted principles and truths, including the links between economics and society, state and citizen, labour and income. As a practice, a modest politics invites participation in the creation of emplaced sustainability and radical democracies.

The path forward will not be easy. It will require transcending widespread currents of pessimism and revanchism. However, through reframing discussion, it is possible to recognize and cultivate emergent, strategic, affective investments in another politics. Fostering new political identities and alliances, a modest politics opens new horizons for the pursuit of just sustainability. "It is imperative", Paulo Freire (1997: 106) wrote, "that we maintain hope even when the harshness of reality may suggest the opposite." In a world rife with conflict and crisis, *Enough!* is a reminder that we can hold an affirmative vision of a liveable future where, despite constraints, all can live well.

NOTES

Introduction

1. Throughout *Enough!*, we use "kyriarchy" to describe unequal and unjust power relations across many domains (Schüssler Fiorenza 1992; Osborne 2015). In some places we complement this with examples and literature on, for example, race, colonialism and gender. Etymologically, kyriarchy is a combination of the Greek terms *kyrios* (which means lord, master or authority) and *arkhia* (which refers to a system of governing, leading or ruling). Scholars using the term recognize many types of hierarchies, including and beyond those configured around class, race, gender, as well as those related to religion, ability, xenophobia, species and so on. Their writing works to shift attention away from particular identities or categories towards understanding broader entanglements of power, structures and systems, and how these shape oppression and inequality.

   Our intention in using "kyriarchy" is not to erase particular inequalities and injustices: they should be named and addressed where we see them! We use the term, instead, as a reminder that an exhaustive list is an impossible task, necessarily partial and sequential, prioritizing some dimensions and terms. We can and should be better at expanding the scope of our attention – scholars writing from disability studies, for example, have shown that this perspective is regularly absent from the listing of injustices in environmental thought (Alaimo 2017). Thinking about intersectionality helps us to understand and analyse multiple identities and their construction, contestation and recreation. Yet, with its emphasis on identity and categories, intersectionality is not synonymous with kyriarchy. The term "kyriarchy" helps us to remember we do not always see all these different forms and that not all injustices have labels and neat categories. We consider the language and analytic of kyriarchy to be rooted in a relational and processual vision of justice, one that provides a sharp and focused object of critique. Thus, to clearly state what we are against with a single term, rather than pursue a more thorough list, we use the term "kyriarchy" to centralize and specify our concern with power, inequality and injustice.
2. The difference between a modern and modest approach to enough for all is usefully captured in the distinction between universal basic services and universal basic income, which we discuss in more detail in Chapter 4.
3. These are composite characters drawn from our research; we use composite characters because we return to Chris and Sipho in the conclusion to imagine possible futures.
4. We hope the reader will forgive our considerable citation of our own work throughout this book! We do so in order to help the interested reader fill gaps and point to empirical examples where we work through many of these arguments in a more concrete fashion. We also hope to help the reader see the bending pathway and emergence of the line of thinking that underpins our argument. Finally, we want to ensure credit to myriad co-authors whose ideas have helped shape our work.
5. We provide more details on modernity and its inverse in the chapter that follows.

6. Imaginaries are abstractions about the world, often implicitly undergirding how people think and construct arguments rather than explicitly stated. Despite their abstraction, imaginaries have profound impacts on beliefs about what is possible and what is desirable. They shape the worlds we build and the politics we practise.
7. Anarchists have a complicated relationship with modernity: while they are best known for their rejection of the state, as Day (2006) convincingly argues, the futures imagined in much anarchist thought rely on modernist ideas of rational humanism and the erasure of inequalities and difference.

1. Polarizing political ecologies of the future

1. For example, in early and ongoing efforts of some degrowth scholars to methodically quantify material flows and determine the limits of ecosystems, we can see elements of a modernist quest to know the world and count it, and scientifically calculate its limits. Such efforts in many ways accord with the modernist efforts to know, control and plan described below. Yet in arguing for degrowth, an undoing of what has been done, such scholars have a fundamentally different view of progress and the ongoing betterment of the human condition that the modernists we examine in more detail below.
2. Really. In October 2019, Philips, Kallis, Robbins and others had an extended conversation on Twitter around the merits of toasters. This was, in a sense, a rather detailed discussion about a kitchen appliance. Yet it also neatly captured the stark contrasts between those seeking to reduce the number of appliances in the typical kitchen, those who believe in building more, better toasters, and those seeking to work beyond such a frame.
3. While a significant body of thought under the banner of ecosocialism relies on modernist frameworks, *not all ecosocialists are modernists* and only a few explicitly label themselves as such. Many advocates of degrowth, for example, position themselves as ecosocialists while selectively adopting and critiquing aspects of modernity. We work to exemplify this point throughout the book.

2. Neither more nor less: cultivating a modest political ecology

1. This call to do political ecology differently is exemplified in the 2022 Dimensions of Political Ecology conference, in which the organizers called for "Radical Imaginings & EcoFuturisms". The call explains, "we do not mean linear progress narratives, simple technofixes for complex crises, or billionaires dreaming up space colonies ... we seek to foreground scholarship, activism, and art that radically envisions different, more hopeful, more abundant futures. Ones that may or may not be known, mapped, or closely planned, but take shape in the praxis of here and now. We ask what comes next? For whom? By whom? With whom?" We write *Enough!* in solidarity with this call.
2. Robbins (2020b) and Jasanoff (2007), for example, suggest the importance of humility in response to ongoing uncertainties. We are largely in agreement with their wider arguments in these works, and hope to offer a term that helps clarify such lines of thinking. We suggest "modest" better approximates the idea of possibilities without undercutting the significance of human agency, and cautiously suggest that our wider line of argument accords with the approach taken by both authors.
3. The term "anticipatory guidance" is used in health sciences to refer to professional recommendations for preventative health; we adopt the phrase with a different intention. We also recognize a wider science of anticipation, but distinguish our thinking here from behaviourist modelling.

3. A modest economy

1. Distributionism as an explicit political economic proposal is rooted in Christian theology and emphasizes the merits of small-scale private property (see Aleman 2015). Other religions have similarly called for the state to redistribute: the Old Testament provided prescriptions to ensure fair shares through the Jubilee, and the Quran, similarly, requires redistribution through the pillar Zakat.

5. Modest livelihoods

1. Our argument here largely accords with Weber's description of the Protestant work ethic, and also draws on Marxist explications of the link between this ethic and capitalism (see Weeks 2011). As we draw on both approaches and note that this ethic has largely been dissociated from its religious connotations, we refer to a "modern work ethic", using language congruent with the wider argument in our book.

REFERENCES

Acosta, A. & J. Cajas-Guijarro 2020. "Ghosts, pluriverse and hopes: from 'development' to post-development". In E. Vivares (ed.), *The Routledge Handbook to Global Political Economy: Conversations and Inquiries*, 292–308. New York: Routledge.

Africa, S. 2022. "South Africa's deadly July 2021 riots may recur if there's no change". *The Conversation*, 9 July. https://theconversation.com/south-africas-deadly-july-2021-riots-may-recur-if-theres-no-change-186397.

Aguilar, F. Jr. 2005. "Excess possibilities? Ethics, populism and community economy: a commentary on JK Gibson-Graham's 'Surplus possibilities: postdevelopment and community economies'". *Singapore Journal of Tropical Geography* 2(1): 27–31.

Alaimo, S. (ed.) 2017. *Disability Studies and the Environmental Humanities*. Lincoln, NE: University of Nebraska Press.

Albert, M. 2017. *Practical Utopia*. Oakland, CA: PM Press.

Alda-Vidal, C., M. Lawhon, D. Iossifova & A. L. Browne (in press). "Living with fragile infrastructure: the gendered labour of preventing, responding to and being impacted by sanitation failures." *Geoforum*.

Aleman, R. (ed.) 2015. *The Hound of Distributism*. New Delhi: ACS Books.

Ali, S. 2007. "Feminism and postcolonial: knowledge/politics: introduction: feminist and postcolonial: challenging knowledge". *Ethnic and Racial Studies* 30(2): 191–212. doi:10.1080/01419870601143877.

Alik-Lagrange, A. & M. Ravallion 2018. "Workfare versus transfers in rural India". *World Development* 112: 244–58. doi:10.1016/j.worlddev.2018.08.008.

Alstott, L. 2004. *No Exit*. New York: Oxford University Press.

Amin, S. 2009. *Eurocentrism*. New York: Monthly Review Press.

Angel, J. & A. Loftus 2019. "With-against-and-beyond the human right to water". *Geoforum* 98: 206–13. doi:10.1016/j.geoforum.2017.05.002.

Archibald, J. 2008. *Indigenous Storywork*. Vancouver, BC: UBC Press.

Armstrong, K. 2000. *The Battle for God*. New York: Knopf.

Arnold, C. 2018. "Money for nothing: the truth about universal basic income". *Nature* 557(7707): 626–8. doi:10.1038/d41586-018-05259-x.

Aronoff, K. *et al.* 2019. *A Planet to Win*. New York: Verso.

Arrighi, G. 1994. *The Long Twentieth Century*. London: Verso.

Asafu-Adjaye, J. *et al.* 2015. "An ecomodernist manifesto". www.ecomodernism.org/.

Bailey, I., A. Gouldson & P. Newell 2011. "Ecological modernisation and the governance of carbon: a critical analysis". *Antipode* 43(3): 682–703. doi:10.1111/j.1467-8330.2011.00880.x.

Ballard, R. 2013. "Geographies of development II: cash transfers and the reinvention of development for the poor". *Progress in Human Geography* 37(6): 811–21. doi:10.1177/0309132512474739.

Barca, S., E. Chertkovskaya & A. Paulsson 2019. "Introduction. The end of political economy as we knew it? From growth realism to nomadic utopianism". In E. Chertkovskaya, A. Paulsson & S. Barca (eds), *Towards a Political Economy of Degrowth*, 1–17. Lanham, MD: Rowman & Littlefield.

Barchiesi, F. 2011. *Precarious Liberation*. New York: SUNY Press.

Barnes, J. 2022. *Staple Security*. Durham, NC: Duke University Press.

Barron, E., L. Hartman & F. Hagemann 2020. "From place to emplacement: the scalar politics of sustainability". *Local Environment* 25(6): 447–62. doi:10.1080/13549839.2020.1768518.

Bastagli, F. *et al.* 2016. *Cash Transfers: What Does the Evidence Say?* London: Overseas Development Institute. https://cdn.odi.org/media/documents/11316.pdf.

Battiste, M. & J. Henderson 2000. *Protecting Indigenous Knowledge and Heritage*. Saskatoon: Purich.

Bawaka Country *et al.* 2016. "Co-becoming Bawaka: towards a relational understanding of place/space". *Progress in Human Geography* 40(4): 455–75. doi:10.1177/0309132515589437.

Beck, U., A. Giddens & S. Lash 1994. *Reflexive Modernization*. Cambridge: Polity.

Beinin, J. 2016. *Workers and Thieves*. Stanford, CA: Stanford University Press.

Bell, B., J. Gaventa & J. Peters (eds) 1990. *Myles Horton and Paulo Freire: We Make the Road by Walking*. Philadelphia, PA: Temple University Press.

Benjamin, R. 2019. *Race After Technology*. Cambridge: Polity.

Berkhout, F., A. Smith & A. Stirling 2004. "Socio-technological regimes and transition contexts". In B. Elzen, F. Geels & K. Green (eds), *System Innovation and the Transition to Sustainability*, 48–75. Cheltenham: Elgar.

Berkes, F. 1999. *Sacred Ecology*. London: Taylor & Francis.

Bhalla, A. (ed.) 1992. *Environment, Employment, and Development*. Geneva: International Labour Organization.

Bhattacharya, T. & L. Vogel 2017. *Social Reproduction Theory*. London: Pluto Press.

Bidadanure, J. 2019. "The political theory of universal basic income". *Annual Review of Political Science* 22: 481–501. doi:10.1146/annurev-polisci-050317-070954.

Bigger, P. 2012. "Red terror on the atmosphere: responses to Wainwright and Mann's 'climate Leviathan'". *Antipode*. https://antipodeonline.org/wp-content/uploads/2012/07/bigger_climate-leviathan.pdf.

Birou, A., P. Henry & J. Schlegel (eds) 1977. *Towards a Re-Definition of Development*. Oxford: Pergamon.

Bix, A. 2000. *Inventing Ourselves Out of Jobs?* Baltimore, MD: Johns Hopkins University Press.

Bledsoe, A., T. McCreary & W. Wright 2022. "Theorizing diverse economies in the context of racial capitalism". *Geoforum* 132: 281–90. doi:10.1016/j.geoforum.2019.07.004.

Bond, P. 2019. "Tokenistic water and neoliberal sanitation in post-apartheid Durban". *Journal of Contemporary African Studies* 37(4): 275–93. doi:10.1080/02589001.2019.1710115.

Borrows, J. 2010. *Drawing Out Law*. Toronto: University of Toronto Press.

Braidotti, R. 2013. *The Posthuman*. Cambridge: Polity.

Braun, B. 2015. "From critique to experiment? Rethinking political ecology for the Anthropocene". In T. Perreault, G. Bridge, J. McCarthy (eds), *The Routledge Handbook of Political Ecology*, 124–36. Abingdon: Routledge.

Braun, B. & S. Whatmore (eds) 2010. *Political Matter*. Minneapolis, MN: University of Minnesota Press.

Bregman, R. 2020a. *Humankind*. London: Bloomsbury.

Bregman, R. 2020b. "The neoliberal era is ending. What comes next?" *The Correspondent*, 14 May. https://thecorrespondent.com/466/the-neoliberal-era-is-ending-what-comes-next.

brown, a. m. 2017. *Emergent Strategy*. Chico, CA: Aka Press.

Brown, M. 1995. *Africa's Choices*. New York: Routledge.

Brownell, E. 2020. *Gone to Ground*. Pittsburgh, PA: University of Pittsburgh Press.

Brulle, R. & R. Dunlap 2021. "A sociological view of the effort to obstruct action on climate change". *Footnotes*. www.asanet.org/sociological-view-effort-obstruct-action-climate-change.

Bryant, R. & S. Bailey 1997. *Third World Political Ecology*. London: Routledge.

Buch-Hansen, H. 2018. "The prerequisites for a degrowth paradigm shift: insights from critical political economy". *Ecological Economics* 146: 157–63. doi:10.1016/j.ecolecon.2017.10.021.

Büchs, M. 2021. "Sustainable welfare: how do universal basic income and universal basic services compare?" *Ecological Economics* 18: 107152. doi:10.1016/j.ecolecon.2021.107152.

Bulkeley, H. & A. Mol 2003. "Participation and environmental governance: consensus, ambivalence and debate". *Environmental Values* 12(2): 143–54. doi:10.3197/096327103129341261.

Burawoy, M. 1985. *The Politics of Production*. London: Verso.

Burnley, M. 2017. "The environmental scandal in Scott Pruitt's backyard". *Politico*, 6 December. www.politico.com/magazine/story/2017/12/06/scott-pruitt-tar-creek-oklahoma-investigation-215854/.

Butler, O. 1993. *Parable of the Sower*. New York: Four Walls Eight Windows.

Butler, O. 1998. *Parable of the Talents*. New York: Seven Stories Press.

Cahill, C. & R. Pain 2019. "Representing slow violence and resistance: on hiding and seeing". *ACME* 18(5): 1054–65. https://acme-journal.org/index.php/acme/article/view/1923.

Calnitsky, D. 2017. "Debating basic income". *Catalyst* 1(3): 63–90. https://catalyst-journal.com/2017/12/debating-basic-income.

Calnitsky, D. & P. Gonalons-Pons 2021. "The impact of an experimental guaranteed income on crime and violence". *Social Problems* 68(3): 778–98. doi:10.1093/socpro/spaa001.

Calzada, I. 2021. "Data co-operatives through data sovereignty". *Smart Cities* 4(3): 1158–72. doi:10.3390/smartcities4030062.

Carson, R. 1962. *Silent Spring*. Boston, MA: Houghton Mifflin.

Casselman, B. 2021. "Pandemic wave of automation may be bad news for workers". *New York Times*, 3 July. www.nytimes.com/2021/07/03/business/economy/automation-workers-robots-pandemic.html.

Castree, N. *et al.* (eds) 2010. *The Point is to Change It*. Oxford: Wiley-Blackwell.

Carter, M. 2015. *Challenging Social Inequality*. Durham, NC: Duke University Press.

Césaire, A. 1939. *Notebook of a Return to the Native Land*. Middletown, CT: Wesleyan University Press.

Chandler, D. (2018). *Ontopolitics in the Anthropocene*. New York: Routledge.

Chakrabarty, D. 2011. "The muddle of modernity". *American Historical Review* 116(3): 663–75. doi:10.1086/ahr.116.3.663.

Charbonnier, P. 2021. *Affluence and Freedom*. Cambridge: Polity.

Checker, M. 2020. *The Sustainability Myth: Environmental Gentrification and the Politics of Justice*. New York: NYU Press.

Clancy, C., F. Cooke & Z. Raw 2021. "Entanglement, autonomy and the co-production of landscapes: relational geographies for free-roaming 'feral' donkeys (Equus asinus) in a rapidly changing world". *Geoforum* 123: 66–77. doi:10.1016/j.geoforum.2021.04.026.

Clark, J. & C. Martin (eds) 2004. *Anarchy, Geography, Modernity*. Lanham, MD: Lexington.

Coelho, A. 2019. "Rutger Bregman: He is saying out loud what the majority of people is thinking". *Basic Income Earth Network*, 9 March. https://basicincome.org/topic/inequality/.

Cohen, D. 2021. "'A marketplace of schools': Race, power, and education reform in the Detroit region". *Urban Geography* 42(8):1170–94. doi:10.1080/02723638.2020.1759015.

Collard, R., J. Dempsey & J. Sundberg 2016. "The Moderns' amnesia in two registers". *Environmental Humanities* 7(1): 227–32. doi:10.1215/22011919-3616425.

Collier, M. *et al.* 2013. "Transitioning to resilience and sustainability in urban communities". *Cities* 32(1): S21–28. doi:10.1016/j.cities.2013.03.010.

Collomb, J. 2019. "Pushing for efficiency: Gifford Pinchot and the first National Parks". *Miranda* 19. doi:10.4000/miranda.20268.

Comaroff, J. & J. Comaroff 1991. *Of Revelation and Revolution*. Chicago, IL: University of Chicago Press.

Comaroff, J. & J. Comaroff 2002. "Alien-nation: zombies, immigrants, and millennial capitalism". *South Atlantic Quarterly* 10(4): 779–805. doi:10.1215/00382876-101-4-779.

Coombes, B., J. Johnson & R. Howitt 2012. "Indigenous geographies I: mere resource conflicts? The complexities in Indigenous land and environmental claims". *Progress in Human Geography* 36(6): 810–21. doi:10.1177/0309132511431410.

Cooper, F. 1996. *Decolonization and African Society*. Cambridge: Cambridge University Press.

Cooper, F. 2005. *Colonialism in Question*. Berkeley, CA: University of California Press.

Coote, A. 2019. "Universal basic income doesn't work. Let's boost the public realm instead". *The Guardian*, 6 May. www.theguardian.com/commentisfree/2019/may/06/universal-basic-income-public-realm-poverty-inequality.

Coote, A. & N. Lawson 2021. "Post-Covid Britain needs a new social guarantee". *The Guardian*, 11 March. www.theguardian.com/commentisfree/2021/mar/11/post-covid-britain-new-social-guarantee-universal-basic-income-pandemic.

Coote, A. & A. Percy 2020. *The Case for Universal Basic Services*. Cambridge: Polity.

Corwin, J. 2018. "'Nothing is useless in nature' Delhi's repair economies and value-creation in an electronics 'waste' sector". *Environment and Planning A* 50(1): 14–30. doi:10.1177/0308518X177390.

Corwin, J. & V. Gidwani 2021. "Repair work as care: on maintaining the planet in the Capitalocene". *Antipode*. Early view. doi:10.1111/anti.12791.

Coulthard, G. 2014. *Red Skin, White Masks*. Minneapolis, MN: University of Minnesota Press.

Cox, S. 2018. *Breaching the Peace*. Vancouver, BC: UBC Press.

Crenshaw, K. 2019. *On Intersectionality*. New York: The New Press.

Creutzig, F. *et al.* 2022. "Demand-side solutions to climate change mitigation consistent with high levels of well-being". *Nature Climate Change* 12(1): 36–46. doi:10.1038/s41558-021-01219-y.

Cronon, W. 1996. "The trouble with wilderness; or, getting back to the wrong nature". In W. Cronon (ed.), *Uncommon Ground*, 69–90. New York: Norton.

Cudd, A. 2006. *Analyzing Oppression*. New York: Oxford University Press.

Curley, A. & S. Smith 2020. "Against colonial grounds: geography on Indigenous lands". *Dialogues in Human Geography* 10(1): 37–40. doi:10.1177/2043820619898900.

Curtis, B. 2002. *The Politics of Population*. Toronto, ON: University of Toronto Press.

Cusset, F. & N. Wedell 2018. *How the World Swung to the Right*. Cambridge, MA: Semiotexte/Smart Art.

D'Alisa, G., F. Demaria & G. Kallis (eds) 2015. *Degrowth*. New York: Routledge.

Daigle, M. 2019. "Tracing the terrain of Indigenous food sovereignties". *Journal of Peasant Studies* 46(2): 297–315. doi:10.1080/03066150.2017.1324423.

Dalla Costa, M. & S. James 1972. *The Power of Women and the Subversion of the Community*. Bristol: Falling Wall Press.

Daly, H. 1977. *Steady-State Economics*. Washington, DC: Island Press.

Davies, A. 2001. "What silence knows-planning, public participation and environmental values". *Environmental Values* 10(1): 77–102. doi:10.3197/096327101129340750.

Davis, A. 2015. *Freedom Is a Constant Struggle*. Chicago, IL: Haymarket.

Davis, H. & Z. Todd 2017. "On the importance of a date, or, decolonizing the Anthropocene". *ACME* 16(4): 761–80. https://acme-journal.org/index.php/acme/article/view/1539.

Davis, M. 2010. "Who will build the ark?" *New Left Review* 61: 26–9.

Dawson, H. 2021. "'Be your own boss': entrepreneurial dreams on the urban margins of South Africa". In W. Monteith, D. Vicol & P. Williams (eds), *Beyond the Wage*, 115–38. Bristol: Bristol University Press.

Day, R. 2006. *Gramsci is Dead*. London: Pluto Press.

de Bont, R. & J. Lachmund 2017. *Spatializing the History of Ecology*. New York: Routledge.

de Sousa Santos, B. 2014. *Epistemologies of the South*. Boulder, CO: Paradigm.

Deighton, A. & E. Dyck 2017. *Managing Madness*. Winnipeg, MB: University of Manitoba Press.

Demaria, F. *et al.* 2013. "What is degrowth? From an activist slogan to a social movement". *Environmental Values* 22(2): 191–215. doi:10.2307/23460978.

Desvallées, L., X. de Sartre & C. Kull 2022. "Epistemic communities in political ecology: critical deconstruction or radical advocacy?" *Journal of Political Ecology* 29(1): 309–40. https://halshs.archives-ouvertes.fr/halshs-03672937.

Dicke, A., N. Safavian & J. Eccles 2019. "Traditional gender role beliefs and career attainment in STEM: A gendered story?". *Frontiers in Psychology* 1053. doi:10.3389/fpsyg.2019.01053.

Dietz, R. & D. O'Neill 2013. *Enough is Enough*. London: Routledge.

Dimensions of Political Ecology 2022. *Guiding Idea: Radical Imaginings & EcoFuturisms*. http://politicalecology.org/2022/theme.

Dixon, C. 2014. *Another Politics*. Oakland, CA: University of California Press.

Durkheim, É. 1997 [1893]. *The Division of Labour in Society*. New York: Free Press.

Earley, L. 2004. *Looking for Longleaf*. Chapel Hill, NC: University of North Carolina Press.

Edwards, G. & H. Bulkeley 2018. "Heterotopia and the urban politics of climate change experimentation". *Environment and Planning D* 36(2): 350–69. doi:10.1177/0263775817747885.

Edwards, P. 2010. *A Vast Machine*. Cambridge, MA: MIT Press.

Eisenstadt, S. 2002. *Multiple Modernities*. Piscataway, NJ: Transaction.

Elhefnawy, N. 2022. *Contextualizing the Great Resignation*. http://dx.doi.org/10.2139/ssrn.4003351.

Eren, A. 2017. "The political ecology of uncertainty: the production of truth by juridical practices in hydropower development". *Journal of Political Ecology* 24(1): 386–405. doi:10.2458/v24i1.20879.

Ernstson, H. & S. Sörlin 2013. "Ecosystem services as technology of globalization: on articulating values in urban nature". *Ecological Economics* 86: 274–84. doi:10.1016/j.ecolecon.2012.09.012.

Escobar, A. 1994. *Encountering Development*. Princeton, NJ: Princeton University Press.

Escobar, A. 2015. "Degrowth, postdevelopment, and transitions: a preliminary conversation". *Sustainability Science* 10(3): 451–62. doi:10.1007/s11625-015-0297-5.

Escobar, A. 2018. *Designs for the Pluriverse*. Durham, NC: Duke University Press.

Esteva, G. & A. Escobar 2017. "Post-Development@25: on 'being stuck' and moving forward, sideways, backward and otherwise". *Third World Quarterly* 38(12): 2559–72. doi:10.1080/01436597.2017.1334545.

Evans, D. & A. Popova 2017. "Cash transfers and temptation goods". *Economic Development and Cultural Change* 65(2): 189–221. doi:10.1086/689575.

Everdell, W. 1997. *The First Moderns*. Chicago, IL: University of Chicago Press.

Fairlie, S. 2009. "A short history of enclosure in Britain. how our land was privatized over five centuries". *The Land* 7: 16–31. www.thelandmagazine.org.uk/articles/short-history-enclosure-britain.

Federici, S. 1975. *Wages Against Housework*. Montpelier: Falling Wall Press.

Federici, S. 2004. *Caliban and the Witch*. New York: Autonomedia.

Federici, S. 2012. *Revolution at Point Zero*. Oakland, CA: PM Press.

Ferguson, J. 2005. "Decomposing modernity: history and hierarchy after development". In A. Loomba *et al.* (eds), *Postcolonial Studies and Beyond*, 166–81. Durham, NC: Duke University Press.

Ferguson, J. 2010. "The uses of neoliberalism". *Antipode* 41(1): 166–84. doi:10.1111/j.1467-8330.2009.00721.x.

Ferguson, J. 2015. *Give a Man a Fish*. Durham, NC: Duke University Press.

Ferguson, J. 2019. "Proletarian politics today: on the perils and possibilities of historical analogy". *Comparative Studies in Society and History* 61: 4–22. doi:10.1017/S0010417518000476.

Ferguson, J. 2021. *Presence and Social Obligation*. Chicago, IL: Prickly Paradigm Press.

Ferguson, J. & T. Li 2018. "Beyond the 'Proper Job': political-economic analysis after the century of labouring man". Working Paper 51, PLAAS, University of the Western Cape, Cape Town. https://repository.uwc.ac.za/bitstream/handle/10566/4538/wp_51_beyond_proper_job_political_economic-analysis_2018.pdf?sequence=1&isAllowed=y.

Fickey, A. & K. Hanrahan 2014. "Moving beyond Neverland: reflecting upon the state of the diverse economies research program and the study of alternative economic spaces". *ACME* 13(2): 394–403. https://acme-journal.org/index.php/acme/article/view/1013.

Fleming, P. 2017. *The Death of Homo Economicus*. London: Pluto Press.

Fletcher, R. & B. Büscher 2020. "Conservation basic income: a non-market mechanism to support convivial conservation". *Biological Conservation* 244: 108520. doi:10.1016/j.biocon.2020.108520.

Flyvbjerg, B. 2005. "Social science that matters". *Foresight Europe* 2: 38–42. https://citeseerx.ist.psu.edu/viewdoc/download?doi=10.1.1.734.4015&rep=rep1&type=pdf.

Forget, E. 2011. "The town with no poverty: the health effects of a Canadian guaranteed annual income field experiment". *Canadian Public Policy* 37(3): 283–305. doi:10.3138/cpp.37.3.283.

Forsyth, T. 2001. *Critical Political Ecology*. London: Routledge.

Fortier, C. 2017. *Unsettling the Commons*. Winnipeg, MB: ARP Books.

Foucault, M. 1972. *The Archaeology of Knowledge*. London: Tavistock.

Foucault, M. 1977. *Discipline and Punish*. New York: Random House.

Fouéré, M. 2014. "Julius Nyerere, Ujamaa, and political morality in contemporary Tanzania". *African Studies Review* 57(1): 1–24.

Fouksman, E. 2021a. "Defending the wage: visions of work and distribution in Namibia". In W. Monteith, D.-O. Vicol & P. Williams (eds), *Beyond the Wage*, 277–94. Bristol: Bristol University Press.

Fouksman, E. 2021b. "Jobs or income guarantees? The politics of universal basic income and cash transfers in Southern Africa". In P. Sloman, D. Vargas & P. Pinto (eds), *Universal Basic Income in Historical Perspective*, 225–45. London: Palgrave Macmillan.

Fourie, E. 2012. "A future for the theory of multiple modernities: insights from the new modernization theory". *Social Science Information*, 51(1): 52–69. https://doi.org/10.1177/0539018411425850.

Frank, A. 1967. *Capitalism and Underdevelopment in Latin America*. New York: Monthly Review Press.

Fraser, N. 2019. *The Old is Dying and the New Cannot be Born*. London: Verso.

Freire, P. 1997. *Pedagogy of the Heart*. New York: Continuum.

Freund, B. 2013. "Labour studies and labour history in South Africa: perspectives from the apartheid era and after". *International Review of Social History* 58(3): 493–519. doi:10.1017/S0020859013000217.

Funiciello, T. 1993. *Tyranny of Kindness*. New York: Atlantic Monthly Press.

Galdino, K. *et al.* 2018. "The informal economy in pan-Africa: review of the literature, themes, questions, and directions for management research". *Africa Journal of Management* 4(3): 225–58. doi:10.1080/23322373.2018.1517542.

Gandhi I. 1972. "Of man and his environment". Plenary session of the United Nations Conference on Human Environment, Stockholm, 14 June.

Gandy, M. 1999. "Rethinking the ecological leviathan: environmental regulation in an age of risk". *Global Environmental Change* 9(1): 59–69. doi:10.1016/S0959-3780(98)00023-5.

George, D. 1967. *Lament for Confederation*. Canadian Broadcasting Corporation, Information Services.

Gergan, M. & T. McCreary 2022. "Disrupting infrastructures of colonial hydro-modernity: Lepcha and Dakelh struggles against temporal and territorial displacements". *Annals of the American Association of Geographers* 112(3): 789–98. doi:10.1080/24694452.2021.1978837.

Gibson-Graham, J. K. 1996. *The End of Capitalism (As We Knew It)*. Oxford: Blackwell.

Gibson-Graham, J. K. 2006. *A Postcapitalist Politics*. Minneapolis, MN: University of Minnesota Press.

Gibson-Graham, J. K. 2007. "Surplus possibilities: post-development and community economies". In A. Ziai (ed.), *Exploring Post-Development*, 145–62. London: Routledge.

Gibson-Graham, J. K. *et al.* 2018. "Cultivating community economies". The Next System Project. https://thenextsystem.org/cultivating-community-economies.

Gidwani, V. & S. Chari 2004. "Geographies of work". *Environment and Planning D* 22(4): 54–84. doi:10.1068/d2204ed.

Gille, Z. 2007. *From the Cult of Waste to the Trash Heap of History*. Bloomington, IN: Indiana University Press.

Gilmore, R. 2022. *Abolition Geography*. London. Verso.

Gilroy, P. 1995. *The Black Atlantic*. Cambridge, MA: Harvard University Press.

Gintilini, U. 2021. "A game changer for social protection: six reflections on COVID-19 and the future of cash transfers". World Bank blogs, 11 January. https://blogs.worldbank.org/developmenttalk/game-changer-social-protection-six-reflections-covid-19-and-future-cash-transfers.

Gneiting, U., N. Lusiani & I. Tamir 2020. *Power, Profits and the Pandemic*. Oxfam Briefing Paper. doi:10.2139/ssrn.3956244.

Goldfischer, E., L. Rice & S. Black 2020. "Obstinate curiosity and situated solidarity in urban political ecology". *Geography Compass* 14(2): e12479. doi:10.1111/gec3.12479.

Goldstein, J. 2018. *Planetary Improvement*. Cambridge, MA: MIT Press.

Golley, F. 1993. *A History of the Ecosystem Concept in Ecology*. New Haven, CT: Yale University Press.

Gómez-Baggethun, E. 2020. "More is more: scaling political ecology within limits to growth". *Political Geography* 76: 102095. doi:10.1016/j.polgeo.2019.102095.

Gorz, A. 1967. *Strategy for Labor*. Boston, MA: Beacon Press.

Gorz, A. 1980. *Ecology as Politics*. Quebec: Black Rose Books.

Gorz, A. 1982. *Farewell to the Working Class*. London: Pluto Press.

Gough, I. 2021. "Move the debate from Universal Basic Income to Universal Basic Services". UNESCO Inclusive Policy Lab. https://en.unesco.org/inclusivepolicylab/analytics/move-debate-universal-basic-income-universal-basic-services.

Graeber, D. 2018. *Bullshit Jobs*. London: Penguin.

Graeber, D. & D. Wengrow 2021. *The Dawn of Everything*. New York: Farrar, Straus & Giroux.

Graham, S. & S. Marvin 2001. *Splintering Urbanism*. London: Routledge.

Grant, R. & M. Oteng-Ababio 2021. "Formalising e-waste in Ghana: an emerging landscape of fragmentation and enduring barriers". *Development Southern Africa* 38(1): 73–86. doi:10.1080/0376835X.2020.1823822.

Grosz, E. 2004. *The Nick of Time*. Durham, NC: Duke University Press.

Grosz, E. 2011. *Becoming Undone*. Durham, NC: Duke University Press.

Grove, R. 1995. *Green Imperialism*. Cambridge: Cambridge University Press.

Grove, K. 2018. *Resilience*. London: Routledge.

Guha, R. & J. Martinez-Alier 2013 [1997]. *Varieties of Environmentalism*. New York: Routledge.

Gulati, R. 2022. "It's not a great resignation – it's a great rethink". *Fortune*. https://fortune.com/2022/03/08/great-resignation-careers-rethink-labor-shortage-pandemic-work-ranjay-gulati/.

Haagh, L. 2019. *The Case for Universal Basic Income*. Cambridge: Polity.

Hacking, I. 1990. *The Taming of Chance*. Cambridge: Cambridge University Press.

Hahnel, R. & E. O. Wright 2016. *Alternatives to Capitalism*. London: Verso.

Hajer, M. 1995. *The Politics of Environmental Discourse*. Oxford: Clarendon Press.

Hanlon, J., A. Barrientos & D. Hulme 2010. *Just Give Money to the Poor*. Sterling, VA: Kumarian Press.

Hannah, M. 2000. *Governmentality and the Mastery of Territory in Nineteenth-Century America*. Cambridge: Cambridge University Press.

Haraway, D. 2016. *Staying with the Trouble*. Durham, NC: Duke University Press.

Hardt, M. & A. Negri 2009. *Commonwealth*. Cambridge, MA: Harvard University Press.

Harman, C. 2004. "The rise of capitalism". *International Socialism* 2(102). http://isj.org.uk/the-rise-of-capitalism/.

Harris, D. 2021. "Storying climate knowledge: notes on experimental political ecology". *Geoforum* 126: 331–9. doi:10.1016/j.geoforum.2021.08.011.

Harvey, D. 1990. *The Condition of Postmodernity*. Oxford: Blackwell.

Harvey, D. 2005. *A Brief History of Neoliberalism*. Oxford: Oxford University Press.

Heimstra, N. 2019. *Detain and Deport*. Athens, GA: University of Georgia Press.

Henderson, M. & M. Lawhon 2021. "What is a 'green city'? Understanding environmental imaginaries of residents in a politically conservative area". *Southeastern Geographer* 61(2): 175–92. doi:10.1353/sgo.2021.0012.

Henrich, J. *et al.* 2005. "'Economic man' in cross-cultural perspective: behavioral experiments in 15 small-scale societies". *Behavioral and Brain Sciences* 28(6): 795–815. doi:10.1017/S0140525X05000142.

Heynen, N. & L. Van Sant 2015. "Political ecologies of activism and direct action politics". In T. Perreault, G. Bridge & J. McCarthy (eds), *The Routledge Handbook of Political Ecology*, 169–78. London: Routledge.

Heynen, N. & M. Ybarra 2021. "On abolition ecologies and making 'freedom as a place'". *Antipode* 53(1): 21–35. doi:10.1111/anti.12666.

Heynen, N. *et al.* (eds) 2007. *Neoliberal Environments*. London: Routledge.

Hickel, J. 2016. "Time for degrowth: to save the planet, we must shrink the economy". *The Conversation*, 23 August. https://theconversation.com/time-for-degrowth-to-save-the-planet-we-must-shrink-the-economy-64195.

Hickel, J. 2020. *Less is More*. London: Penguin.

Hill, H. 1996. "The problem of race in American labor history". *Reviews in American History* 24(2): 189–208. doi:10.1353/rah.1996.0037.

Hirsch, F. 1976. *Social Limits to Growth*. Cambridge, MA: Harvard University Press.

Ho, P. 2003. "Mao's war against nature? The environmental impact of the grain-first campaign in China". *China Journal* 50: 37–59. doi:10.2307/3182245.

Hobsbawm, E. & T. Ranger (eds) 1983. *The Invention of Tradition*. Cambridge: Cambridge University Press.

Hochschild, A. 2016. *Strangers in their Own Land*. New York: New Press.

Holling, C. 1973. "Resilience and stability of ecological systems". *Annual Review of Ecology and Systematics* 4: 1–23. doi:10.1146/annurev.es.04.110173.000245.

Holloway, J. 2002. *Change the World Without Taking Power*. London: Pluto Press.

Holloway, S. & P. Kirby 2020. "Neoliberalising education: new geographies of private tuition, class privilege, and minority ethnic advancement". *Antipode* 52(1): 164–84.

Honwana, A. 2012. *The Time of Youth*. Sterling, VA: Kumarian Press.

hooks, b. 2015. *Feminism is for Everybody*. New York: Routledge.

Huber, M. 2018. "Review of 'This Changes Everything: Capitalism vs. the Climate' by Naomi Klein". *Antipode*. https://antipodeonline.org/2018/12/17/this-changes-everything-capitalism-vs-the-climate.

Huber, M. 2021. "The case for socialist modernism". *Political Geography* 87: 102352.

James, C., R. Dunayevskaya & G. Boggs 2013. *State Capitalism and World Revolution*. Oakland, CA: PM Press.

Jameson, F. 1991. *Postmodernism, or, The Cultural Logic of Late Capitalism*. Durham, NC: Duke University Press.

Jasanoff, S. 2007. "Technologies of humility". *Nature* 450: 33. doi:10.1038/450033a.

Jeffrey, C. & J. Dyson 2021. "Geographies of the future: prefigurative politics". *Progress in Human Geography* 45(4): 641–58. doi:10.1177/0309132520926569.

Jessop. B. 1994. "The transition to post-Fordism and the Schumpeterian workfare state". In R. Burrows & B. Loader (eds), *Towards a Post-Fordist Welfare State?*, 13–37. London: Routledge.

Jessop, B. 2007. *State Power*. Cambridge: Polity.

Johnson, J. & B. Murton 2007. "Re/placing native science: Indigenous voices in contemporary constructions of nature". *Geographical Research* 14(2): 121–9. doi:10.1111/j.1745-5871.2007 00442.x.

Jon, I. 2021. *Cities in the Anthropocene*. London: Pluto Press.

Jones, C. & T. McCreary 2022. "Zombie automobility". *Mobilities* 17(1): 19–36. doi:10.1080/17450101.2021.1940245.

Jones, D. 2010. *The Racial Discourses of Life Philosophy*. New York: Columbia University Press.

Joseph-Gabriel, A. 2019. *Reimagining Liberation*. Urbana-Champaign, IL: University of Illinois Press.

Jungell-Michelsson, J. & P. Heikkurinen 2022. "Sufficiency: a systematic literature review". *Ecological Economics* 195: 107380. doi:10.1016/j.ecolecon.2022.107380.

Kahneman, D. *et al.* 2006. "Would you be happier if you were richer? A focusing illusion". *Science* 312(5782): 1908–10. doi:10.1126/science.1129688.

Kallis, G. 2018. *Degrowth*. Newcastle upon Tyne: Agenda Publishing.

Kallis, G. 2019. *Limits*. Stanford, CA: Stanford University Press.

Kallis, G. & H. March 2015. "Imaginaries of hope: the utopianism of degrowth". *Annals of the Association of American Geographers* 105(2): 360–8. doi:10.1080/00045608.2014.973803.

Kallis, G. *et al.* 2020. *The Case for Degrowth*. Cambridge: Polity.

Kapoor, I. 2008. *The Postcolonial Politics of Development*. London: Routledge.

Kelley, R. 2002. *The Black Radical Imagination*. Boston: Beacon Press.

Kerschner, C. *et al.* 2018. "Degrowth and technology: towards feasible, viable, appropriate and convivial imaginaries". *Journal of Cleaner Production* 197(2): 1619–36. doi:10.1016/j.jclepro.2018.07.147.

Khasnabish, A. 2010. *Zapatistas*. Halifax, NS: Fernwood.

Kimmerer, R. 2013. *Braiding Sweetgrass*. Minneapolis, MN: Milkweed.

King, M. 2010 [1968]. *Where Do We Go from Here: Chaos or Community?* Boston, MA: Beacon Press.

King, T. 2019. *The Black Shoals*. Durham, NC: Duke University Press.

Kipnis, A. 2008. "Audit cultures: neoliberal governmentality, socialist legacy, or technologies of governing?" *American Ethnologist* 35(2): 275–89. doi:10.1111/j.1548-1425.2008.00034.x.

Kitchin, R. 2014. *The Data Revolution*. Los Angeles, CA: Sage.

Kitchin, R. & M. Dodge 2011. *Code/Space*. Cambridge, MA: MIT Press.

Klein, E. & E. Fouksman 2022. "Reparations as a rightful share: from universalism to redress in distributive justice". *Development and Change* 53(1): 31–57. doi:10.1111/dech.12695.

Klein, N. 2014. *This Changes Everything*. New York: Simon & Schuster.

Koch, M. 2022. "Social policy without growth: moving towards sustainable welfare states". *Social Policy and Society* 21(3): 447–59. doi:10.1017/S1474746421000361.

Kohler, R. 2002. *Landscapes and Labscapes: Exploring the Lab-Field Border in Biology*. Chicago, IL: University of Chicago Press.

Kosek, J. 2006. *Understories*. Durham, NC: Duke University Press.

Kothari, A. *et al.* (eds) 2019. *Pluriverse*. New Delhi: Tulika Books/Authorsupfront.

Kovach, M. 2021. *Indigenous Methodologies*. Toronto, ON: University of Toronto Press.

Kovel, J. 2002. *The Enemy of Nature*. New York: Zed Books.

Kurlansky, M. 2005. *1968: The Year that Rocked the World*. New York: Vintage.

Kymlicka, W. 1995. *Multicultural Citizenship*. Oxford: Oxford University Press.

Laclau, E. & C. Mouffe 1985. *Hegemony and Socialist Strategy*. London: Verso.

Lafargue, P. 1907 [1883]. *The Right to be Lazy*. Chicago, IL: Charles H. Kerr & Co.

LaFreniere, G. 1990. "Rousseau and the European roots of environmentalism". *Environmental History Review* 14(4): 41–72. doi:10.2307/3984813.

Larsen, S. & J. Johnson 2017. *Being Together in Place*. Minneapolis, MN: University of Minnesota Press.

Latham, J. 2000. "There's enough food for everyone, but the poor can't afford to buy it". *Nature* 404(6775): 222. doi:10.1038/35005264.

Latour, B. 1994. *We Have Never Been Modern*. Cambridge, MA: Harvard University Press.

Lave, R. 2012. *Fields and Streams: Stream Restoration, Neoliberalism, and the Future of Environmental Science*. Athens, GA: UGA Press.

Lawhon, M. 2021. "A Political Ecology of Environmental Solutions". Open Education Resource. https://docs.google.com/document/d/1HGOpBP7JKpw7QDfwykDEUI20UepVZ2fPJMG3snoX2kY/edit.

Lawhon, M. & M. Chion 2012. "Rooted cosmopolitanism: Spaces of multiplicity in Cusco, Peru". *International Journal of Urban and Regional Research* 36(3): 539–53. doi:10.1111/j.1468-2427.2011.01079.x.

Lawhon, M. & A. Makina 2015. "Assessing local discourses on water in a South African newspaper". *Local Environment* 22(2): 240–55. doi:10.1080/13549839.2016.1188064.

Lawhon, M. & T. McCreary 2020. "Beyond jobs vs environment: on the potential of universal basic income to reconfigure environmental politics". *Antipode* 52(2): 452–74. doi:10.1111/anti.12604.

Lawhon, M. & T. McCreary 2023. "Making UBI radical: on the potential for a universal basic income to underwrite transformative change". *Economy and Society* 52(2): 349–72.

Lawhon, M., N. Millington & K. Stokes 2018a. "A labour question for the 21st century: perpetuating the work ethic in the absence of jobs in South Africa's waste sector". *Journal of Southern African Studies* 44(6): 1115–31. doi:10.1080/03057070.2018.1528764.

Lawhon, M., A. Makina & G. Nakyagaba (2023a). "Infrastructure beyond the modern ideal: thinking through heterogeneity, serendipity and autonomy in African cities". In M. Kaika *et al.* (eds), *Turning Up the Heat*, 186–204. Manchester: Manchester University Press.

Lawhon, M., G. Nakyagaba & T. Karpouzoglou (2023b). "Towards a modest imaginary? Sanitation in Kampala beyond the modern infrastructure ideal". *Urban Studies* 60(1): 146–65. doi:10.1177/00420980211064519.

Lawhon, M. & M. Patel 2013. "Scalar politics and local sustainability: Rethinking governance and justice in an era of political and environmental change". *Environment and Planning C: Government and Policy* 36(1): 1048–62.

Lawhon, M. *et al.* 2018b. "Thinking through heterogeneous infrastructure configurations". *Urban Studies* 55(4): 720–32. doi:10.1177/0042098017720149.

Lawhon, M. *et al.* 2020. *Making Urban Theory*. Abingdon: Routledge.

Leach, M., I. Scoones & A. Stirling 2007. "Pathways to sustainability: an overview of the STEPS Centre approach". Brighton, STEPS Centre. https://steps-centre.org/publication/pathways-to-sustainability-an-overview-of-the-steps-centre-approach/.

Lefebvre, H. 1969. *The Explosion*. New York: Monthly Review Press.

Levien, M., M. Watts & H. Yan 2018. "Agrarian Marxism". *Journal of Peasant Studies* 45(5/6): 853–83. doi:10.1080/03066150.2018.1534101.

Lewis, A. 2015. *This Changes Everything*. Louverture Films.

Lewis, K. & W. Stronge 2021. *Overtime*. London: Verso.

Li, M. 2009. "Capitalism, climate change and the transition to sustainability: alternative scenarios for the US, China and the world". *Development and Change* 40(6): 1039–61. doi:10.1111/j.1467-7660.2009.01611.x.

Li, T. 2007. *The Will to Improve*. Durham, NC: Duke University Press.

Lin, L. & D. Gullotta 2021. "Disarticulating Gingnian: Chinese youth beyond 'rising tides' and 'lying flat'". *Made in China Journal* 6(3): 20–30. doi:10.3316/informit.376044958369310.

Lindell, I. 2010. "Informality and collective organising: identities, alliances and transnational activism in Africa". *Third World Quarterly* 31(2): 207–22. doi:10.1080/01436591003711959.

Livingstone, D. & C. Withers 1999. *Geography and Enlightenment*. Chicago, IL: University of Chicago Press.

Loftus, A. 2020. "Political ecology II: whither the State?" *Progress in Human Geography* 44(1): 139–49. doi:10.1177/030913251880342.

Lombardozzi, L. & F. Pitts 2020. "Social form, social reproduction and social policy: basic income, basic services, basic infrastructure". *Capital & Class* 44(4): 573–94. doi:10.1177/0309816819873323.

Lorimer, J. 2020. *The Probiotic Planet*. Minneapolis, MN: University of Minnesota Press.

Lotringer, S. & C. Marazzi (eds) 2007. *Autonomia*. Los Angeles, CA: Semiotexte.

Lowman, E. & A. Barker 2015. *Settler*. Halifax, NS: Fernwood.

Lowrey, A. 2022. "Generation Z doesn't remember when America worked". *The Atlantic*, 14 August. www.theatlantic.com/ideas/archive/2022/08/progressive-gen-z-politics-voting/671130/.

Luque-Lora, R. 2021. "Engaging imaginaries, rejecting utopias: the case for technological progress and political realism to sustain material wellbeing". *Political Geography* 87: 102358. doi:10.1016/j.polgeo.2021.102358.

Lushaba, L. 2009. *Development as Modernity, Modernity as Development*. Dakar: Codesria.

Luthra, A. 2022. "Complexity and heterogeneity in the informal economy of waste: problems and prospects for organising and formalising". In F. Nunan, C. Barnes & S. Krishnamurthy (eds), *Routledge Handbook on Livelihoods in the Global South*, 324–34. Abingdon: Routledge.

Makhulu, A. 2012. "The conditions for after work: financialization and informalization in posttransition South Africa". *PMLA* 127(4): 782–99. doi:10.1632/pmla.2012.124.4.782.

Makhulu, A. 2016. "A brief history of the social wage: welfare before and after racial Fordism". *South Atlantic Quarterly* 115(1): 113–24. doi:10.1215/00382876-3424775.

Makina, A. & M. Lawhon 2023. "Permission to appropriate: waste pickers' 'guidelines' for contesting and consolidating claims to waste on the streets of South Africa". Geoforum 137: 52–60. doi:10.1016./j.geoforum.2022.10.003.

Mamdani, M. 2012. *Define and Rule: Native as Political Identity*. Cambridge, MA: Harvard University Press.

Mann, G. & J. Wainwright 2018. *Climate Leviathan*. London: Verso.

Marangos, J. 2004. "Social dividend versus basic income guarantee in market socialism". *International Journal of Political Economy* 34(3): 20–40. doi:10.1080/08911916.2004.11042930.

Marlon, J. *et al.* 2022. "Yale climate opinion maps 2021". Yale Program on Climate Change Communication. Visualizations & Data. https://climatecommunication.yale.edu/visualizations-data/ycom-us/.

Marris, E. 2013. *Rambunctious Garden*. New York: Bloomsbury.

Martin, D. & J. Pierce 2023. *How to Think about Cities*. Cambridge: Polity.

Martin, L. & A. Secor 2014. "Towards a post-mathematical topology". *Progress in Human Geography* 38(3): 420–38. doi:10.1177/0309132513508209.

Martinez-Alier, J. 2002. *The Environmentalism of the Poor*. Cheltenham: Elgar.

Marx, K. 1993. *Grundrisse*. Harmondsworth: Penguin.

Massey, D. 2004. "Geographies of responsibility". Geografiska Annaler: *Series B* 86(1): 5–18. doi:10.1111/j.0435-3684.2004.00150.x.

Mastini, R., G. Kallis & J. Hickel 2021. "A green new deal without growth?" *Ecological Economics* 179: 106832. doi:10.1016/j.ecolecon.2020.106832.

Matthaei, J. 1984. "Rethinking scarcity: neoclassicism, neoMalthusianism, and neo-Marxism". *Review of Radical Political Economics* 16(2/3): 81–94. doi:10.1177/048661348401600204.

May, J. *et al.* 2019. "Welfare convergence, bureaucracy, and moral distancing at the food bank". *Antipode* 51(4): 1251–75. doi:10.1111/anti.12531.

Mbembe, A. 2019. *Necropolitics*. Durham, NC: Duke University Press.

McCarthy, J. 2002. "First world political ecology: lessons from the Wise Use movement". *Environment and Planning A* 34(7): 1281–302. doi:10.1068/a3526.

McCarthy, T. 2004. "Coming to terms with our past, part II: on the morality and politics of reparations for slavery". *Political Theory* 32(6): 750–72. doi:10.1177/0090591704268924.

McCreary, T., R. Basu & A. Godlewska 2013. "Critical geographies of education: introduction to the special issue". *The Canadian Geographer* 57(3): 255–9. doi:10.1111/cag.12031.

McCreary, T., S. Mills & A. St-Amand 2016. "Lands and resources for jobs: how Aboriginal peoples strategically use environmental assessments to advance community employment aims". *Canadian Public Policy* 42(2): 212–23. doi:10.3138/cpp.2015-061.

Mcdowell, L. 2004. "Work, workfare, work/life balance and an ethic of care". *Progress in Human Geography* 28(2): 145–63. doi:10.1191/0309132504ph478oa.

McLaren, D. & J. Agyeman 2017. *Sharing Cities*. Cambridge, MA: MIT Press.

Meadows, D. *et al.* 1972. *The Limits to Growth*. New York: Universe Books.

Mehta, L. (ed.) 2010. *The Limits to Scarcity*. London: Earthscan.

Mehta, L. & W. Harcourt 2021. "Beyond limits and scarcity: feminist and decolonial contributions to degrowth". *Political Geography* 89: 102411. doi:10.1016/j.polgeo.2021.102411.

Mehta, L., H. Adam & S. Srivastava 2022. *The Politics of Climate Change and Uncertainty in India*. Abingdon: Routledge.

Mehta L., M. Leach & I. Scoones 2001. "Environmental governance in an uncertain world". *IDS Bulletin* 32(4): 1–9. https://opendocs.ids.ac.uk/opendocs/bitstream/handle/20.500.12413/8744/IDSB_32_4_10.1111-j.1759-5436.2001.mp3 2004001.x.pdf?sequence=1.

Merchant, C. 1980. *The Death of Nature*. San Francisco, CA: Harper & Row.

Mies, M. & V. Shiva 1993. *Ecofeminism*. London: Zed Books.

Mignolo, W. 2007. "Delinking: the rhetoric of modernity, the logic of coloniality and the grammar of de-coloniality". *Cultural Studies* 21(2/3): 449–514. doi:10.1080/09502380601162647.

Mignolo, W. 2011. *The Darker Side of Western Modernity*. Durham, NC: Duke University Press.

Millar, K. 2008. "Making trash into treasure: struggles for autonomy on a Brazilian garbage dump". *Anthropology of Work Review* 29(2): 25–34. doi:10.1111/j.1548-1417.2008.00011.x.

Miller, A. *et al.* 2019. *Community Energy in Western Canada*. Edmonton, AB: Future Energy Systems, University of Alberta. www.futureenergysystems.ca/public/download/files/90232.

Millward-Hopkins, J. *et al.* 2020. "Providing decent living with minimum energy: a global scenario". *Global Environmental Change* 65: 102168. doi:10.1016/j.gloenvcha.2020.102168.

Miraftab, F. 2004. "Neoliberalism and casualization of public sector services: the case of waste collection services in Cape Town, South Africa". *International Journal of Urban and Regional Research* 28(4): 874–92. doi:10.1111/j.0309-1317.2004.00557.x.

Mitchell, D. 2003. *The Right to the City*. New York: Guilford.

Mohai, P., D. Pellow & J. Roberts 2009. "Environmental justice". *Annual Review of Environment and Resources* 34(1): 405–30. doi:10.1146/annurev-environ-082508-094348.

Mohan, D., R. Sekhani & A. Kaushik 2018. "On the ways of knowing and understanding informality". *Asian Journal of Public Affairs* 10(2): 71–9. doi:10.18003/ajpa.20184.

Monbiot, G. 2022. *Regenesis*. London: Penguin.

Monteith, W., D. Vicol & P. Williams 2021. "Introduction: work beyond the wage". In W. Monteith, D. Vicol & P. Williams (eds), *Work Beyond the Wage*: 1–20. Bristol: Bristol University Press.

Moolakkattu, J. 2010. "Gandhi as a human ecologist". *Journal of Human Ecology* 29(3): 151–8. doi:10.1080/09709274.2010.11906257.

Moore, J. 2016. *Anthropocene or Capitalocene?* Oakland, CA: PM Press.

Morales, M., L. Harris & G. Öberg 2014. "Citizenshit: the right to flush and the urban sanitation imaginary". *Environment and Planning A* 46(12): 2816–33. doi:10.1068/a130331p.

Moten, F. 2017. *Black and Blur*. Durham, NC: Duke University Press.

Mouffe, C. 2014. "Democratic politics and conflict: an agonistic approach". In M. Lakitsch (ed.), *Political Power Reconsidered*, 17–29. Zurich: Lit Verlag.

Munck, R. 2013. "The precariat: a view from the south". *Third World Quarterly* 34(5): 747–62. doi:10.1080/01436597.2013.800751.

Mwangi, O. 2007. "Hydropolitics, ecocide and human security in Lesotho: a case study of the Lesotho Highlands Water project". *Journal of Southern African Studies* 33(1): 3–17. doi:10.1080/03057070601136509.

Nakyagaba, G. *et al.* 2021. "Power, politics and a poo pump: contestation over legitimacy, access and benefits of sanitation technology in Kampala". *Singapore Journal of Tropical Geography* 42(3): 415–30. doi:10.1111/sjtg.12381.

Naylor, L. 2017. "Reframing autonomy in political geography: a feminist geopolitics of autonomous resistance". *Political Geography* 58: 24–35. doi:10.1016/j.polgeo.2017.01.001.

Nelson, S. 2014. "Resilience and the neoliberal counter-revolution: from ecologies of control to production of the common". *Resilience* 2(1): 1–17. doi:10.1080/21693293.2014.872456.

Nguyen, N., D. Cohen & A. Huff 2017. "Catching the bus: a call for critical geographies of education". *Geography Compass* 11(8): e12323. doi:10.1111/gec3.12323.

Nirmal, P. & D. Rocheleau 2019. "Decolonizing degrowth in the post-development convergence: questions, experiences, and proposals from two Indigenous territories". *Environment and Planning E* 2(3): 465–92. doi:10.1177/2514848618819478.

Nixon, R. 2011. *Slow Violence and the Environmentalism of the Poor*. Cambridge, MA: Harvard University Press.

Nolan, R. 2018. "Tar Creek: The Quapaw Tribe, the EPA, and tribal self-determination, 1980–2010". *Wicazo Sa Review* 33(1): 70–86. https://muse.jhu.edu/article/723069.

Norgaard, R. 1994. *Development Betrayed*. New York: Routledge.

Nyerere, J. 1968. *Freedom and Socialism: Uhuru Na Ujamaa; a Selection from Writings and Speeches, 1965–1967*. Oxford: Oxford University Press.

O'Brien, P., P. O'Neill & A. Pike 2019. "Funding, financing and governing urban infrastructures". *Urban Studies* 56(7): 1291–303. doi:10.1177/0042098018824014.

O'Connor, J. 1991. "On the two contradictions of capitalism". *Capitalism Nature Socialism* 2(3): 107–9. doi:10.1080/10455759109358463.

O'Hare, P. 2019. "'The landfill has always borne fruit': precarity, formalisation and dispossession among Uruguay's waste pickers". *Dialectical Anthropology* 43(1): 31–44. doi:10.1007/s10624-018-9533-6.

Oreskes, N. & E. Conway 2010. *Merchants of Doubt*. New York: Bloomsbury.

Osborne, N. 2015. "Intersectionality and kyriarchy: a framework for approaching power and social justice in planning and climate change adaptation". *Planning Theory* 14(2): 130–51. doi:10.1177/1473095213516443.

Oswin, N. 2019a. "Enough". In T. Jazeel *et al.* (eds), *Keywords in Radical Geography: Antipode at 50*, 113–17. Chichester: Wiley.

Oswin, N. 2019b. "Enough". Panel with M. Daigle *et al.* American Association of Geographers Annual Meeting, Washington DC, 3 April. https://aag.secure-abstracts.com/AAG%20Annual%20Meeting%202019/sessions-gallery/22638.

Owusu, F. 2020. "Livelihoods". In R. Kitchen & N. Thrift (eds), *International Encyclopedia of Human Geography*, 2nd edn, 193–8. Amsterdam: Elsevier.

Owusu-Addo, E., A. Renzaho & B. Smith 2018. "The impact of cash transfers on social determinants of health and health inequalities in sub-Saharan Africa: a systematic review". *Health Policy and Planning* 33(5): 675–96. doi:10.1093/heapol/czy020.

Özden-Schilling, T. (in press). *Science, Survivance, and the War in the Woods*. Durham, NC: Duke University Press.

Palmer, D. & F. Winiger 2019. "Neo-socialist governmentality: managing freedom in the People's Republic of China". *Economy and Society* 48(4): 554–78. doi:10.1080/03085147.2019.1672424.

Parsons, A. 2018. *From Asylum to Prison*. Chapel Hill, NC: University of North Carolina Press.

Paulson, S. 2021. "Decolonizing technology and political ecology futures". *Political Geography* 88: 102369. doi:10.1016/j.polgeo.2021.102369.

Paulson, S. & L. Gezon (eds) 2005. *Political Ecology across Spaces, Scales, and Social Groups*. Piscataway, NJ: Rutgers University Press.

Pearsall, H. *et al.* 2022. "Building knowledge infrastructure for diverse stakeholders to scale up co-production equitably". *Current Opinion in Environmental Sustainability* 54: 101156. doi:10.1016/j.cosust.2022.101156.

Peck, J. 2001. *Workfare States*. New York: Guilford.

Pellow, D. 2017. *What is Critical Environmental Justice?* Cambridge: Polity.

Pierce, J. 2022. "How can we share space? Ontologies of spatial pluralism in Lefebvre, Butler, and Massey". *Space and Culture* 25(1): 20–32. doi:10.1177/1206331219863314.

Pierce, J. & O. Williams 2016. "Against power? Distinguishing between acquisitive resistance and subversion". Geografiska Annaler: *Series B* 98(3): 171–88. doi:10.1111/geob.12098.

Pierce, J., M. Lawhon & T. McCreary 2019. "From precarious work to obsolete labour? Implications of technological disemployment for geographical scholarship". *Geografiska Annaler: Series B* 101(2): 84–101. doi:10.1080/04353684.2018.1544467.

Piven, F. & R. Cloward 1971. *Regulating the Poor*. New York: Pantheon.

Polanyi, K. 1971. *Primitive, Archaic, and Modern Economies*. Boston, MA: Beacon Press.

Polanyi, K. 2001 [1944]. *The Great Transformation*. Boston, MA: Beacon Press.

Ponder, C. 2021. "Spatializing the municipal bond market: urban resilience under racial capitalism". *Annals of the American Association of Geographers* 111(7): 2112–29. doi:10.1080/24694452.2020.1866487.

Poppendieck, J. 1998. *Sweet Charity?* New York: Viking.

Popper, K. 1978. "Three Worlds". The Tanner Lecture on Human Values, University of Michigan, Ann Arbor, 7 April.

Portes, J., H. Reed & A. Percy 2017. *Social Prosperity for the Future: A Proposal for Universal Basic Services*. Social Prosperity Network Report. London: Institute for Global Prosperity, University College London. https://discovery.ucl.ac.uk/id/eprint/10080175/1/universal_basic_services_-_the_institute_for_global_prosperity_.pdf.

Poulantzas, N. 1978. *State, Power, Socialism*. London: Verso.

Poulsen, C. 1984. *The English Rebels*. London: Journeyman.

Press, A. 2022. "Gig labor is impoverishing workers". Jacobin. https://jacobin.com/2022/06/gig-workers-report-economic-precarity-independent-contracting-misclassification.

Prudham, S. 2007. "Sustaining sustained yield: class, politics, and post-war forest regulation in British Columbia". *Environment and Planning D* 25(2): 258–83. doi:10.1068/d2104.

Pryde, P. 1991. *Environmental Management in the Soviet Union*. New York: Cambridge University Press.

Puar, J. 2017. *The Right to Maim*. Durham, NC: Duke University Press.

Puig de la Bellacasa, M. 2017. *Matters of Care*. Minneapolis, MN: University of Minnesota Press.

Purcell, M. 2014. "Possible worlds: Henri Lefebvre and the right to the city". *Journal of Urban Affairs* 36(1): 141–54. doi:10.1111/juaf.12034.

Raby, M. 2017. *American Tropics*. Columbia, SC: UNC Press.

Rajagopal, B. 2003. *International Law from Below*. Cambridge: Cambridge University Press.

Ramirez, M. 2020. "Take the houses back/take the land back: Black and Indigenous urban futures in Oakland". *Urban Geography* 41(5): 682–893. doi:10.1080/02723638.2020.1736440.

Rangarajan, M. 2006. "Ideology, the environment and policy: Indira Gandhi". *India International Centre Quarterly* 33(1): 50–64. doi:10.4103/0972-4923.65175.

Ravi R. 2017. "A history of environmental justice in India". *Environmental Justice* 7(5): 117–21. doi:10.1089/env.2014.7501.

Rebouças, R. & A. Soares 2021. "Voluntary simplicity: a literature review and research agenda". *International Journal of Consumer Studies* 45(3): 303–19. doi:10.1111/ijcs.12621.

Redclift, M. 1992. "The meaning of sustainable development". *Geoforum* 23(3): 395–403. doi:10.1016/0016-7185(92)90050-E.

Reed, D. 1996. *Structural Adjustment, the Environment and Sustainable Development*. London: Routledge.

Reno, J. 2016. *Waste Away*. Los Angeles, CA: University of California Press.

Robbins, P. 2015. "The trickster science". In T. Perreault, G. Bridge & J. McCarthy (eds), *Routledge Handbook of Political Ecology*, 89–110. Abingdon: Routledge.

Robbins, P. 2020a. "Is less more … or is more less? Scaling the political ecologies of the future". *Political Geography* 76: 102018. doi:10.1016/j.polgeo.2019.04.010.

Robbins, P. 2020b. "Revolutionary humility: response to Gómez-Baggethun". *Political Geography* 76: 102099. doi:10.1016/j.polgeo.2019.102099.

Robbins, P. & S. Smith 2017. "Baby bust: towards political demography". *Progress in Human Geography* 41(2): 199–219. doi:10.1177/0309132516633321.

Robertson, M. 2006. "The nature that capital can see: science, state, and market in the commodification of ecosystem services". *Environment and Planning D* 24(3): 367–87. doi:10.1068/d3304.

Robertson, M. 2012. "Measurement and alienation: making a world of ecosystem services". *Transactions of the Institute of British Geographers* 37(3): 386–401. doi:10.1111/j.1475-5661.2011.00476.x.

Robinson, C. 1983. *Black Marxism*. London: Zed Press.

Rocheleau, D. 2008. "Political ecology in the key of policy: from chains of explanation to webs of relation". *Geoforum* 39(2): 716–27. doi:10.1016/j.geoforum.2007.02.005.

Rocheleau, D., B. Thomas-Slayter & E. Wangari 1996. "Gender and environment: a feminist political ecology perspective". In D. Rocheleau, B. Thomas-Slayter & E. Wangari (eds), *Feminist Political Ecology*, 3–23. London: Routledge.

Rodney, W. 1972. *How Europe Underdeveloped Africa*. London: Bogle-L'Ouverture Publications.

Roosevelt, T. 1903. "Remarks at a Meeting of the Society of American Foresters at the Residence of Mr. Gifford Pinchot, Washington, DC, 26 March 1903". www.presidency.ucsb.edu/documents/remarks-meeting-the-society-american-foresters-the-residence-mr-gifford-pinchot.

Rose, N. 1999. *Powers of Freedom*. New York: Cambridge University Press.

Rose, N. & P. Miller 2010. "Political power beyond the state: problematics of government". *British Journal of Sociology* 43(2): 271–303. doi:10.2307/591464.

Rostow, W. 1960. *The Stages of Economic Growth*. Cambridge: Cambridge University Press.

Roy, A. 2010. *Poverty Capital*. London: Routledge.

Ruckert, A., C. Huynh & R. Labonté 2018. "Reducing health inequities: is universal basic income the way forward?" *Journal of Public Health* 40(1): 3–7. doi:10.1093/pubmed/fdx006.

Saad-Filho, A. 2020. "From COVID-19 to the end of neoliberalism". *Critical Sociology* 46 (4/5): 477–85. doi:10.1177/0896920520929966.

Sachs, A. 2013. *Arcadian America*. New Haven, CT: Yale University Press.

Sachs, W. (ed.) 1992. *The Development Dictionary*. London: Zed Books.

Samantroy, E. & S. Nandi (eds) 2022. *Gender, Unpaid Work and Care in India.* Abingdon: Routledge.

Samers, M. 2005. "The myopia of 'diverse economies', or a critique of the 'informal economy'". *Antipode* 37(5): 875–86. doi:10.1111/j.0066-4812.2005.00537.x.

Sandoval, C. 2000. *Methodology of the Oppressed.* Minneapolis, MN: University of Minnesota Press.

Satterthwaite, D. *et al.* 2019. *Untreated and Unsafe: Solving the Urban Sanitation Crisis in the Global South.* Working Paper. Washington, DC: World Resources Institute.

Sayre, N. 2017. *The Politics of Scale.* Chicago, IL: University of Chicago Press.

Schüssler Fiorenza, E. 1992. *But She Said.* Boston, MA: Beacon Press.

Schneider, N. 2022. "Governable stacks against digital coloniariple*tripleC* 20(1): 19–36. doi:10.31269/triplec.v20i1.1281.

Schreiner, M. 2001. "Informal finance and the design of microfinance". *Development in Practice* 11(5): 637–40. doi:10.1080/09614520120085386.

Schumacher, E. 2011 [1973]. *Small is Beautiful.* London: Random House.

Schuppert, F. 2019. "Green New Deal: universal basic income could make green transition feasible". *The Conversation.* https://theconversation.com/green-new-deal-universal-basic-income-could-make-green-transition-feasible.

Scoones, I., M. Leach & P. Newell 2015. *The Politics of Green Transformations.* Abingdon: Routledge.

Scott, J. 1985. *Weapons of the Weak: Everyday Forms of Peasant Resistance.* New Haven, CT: Yale University Press.

Scott, J. 2012. *Two Cheers for Anarchism.* Princeton, NJ: Princeton University Press.

Seager, J. 2003. "Rachel Carson died of breast cancer: the coming of age of feminist environmentalism". *Signs* 28(3): 945–72. doi:10.1086/345456.

Seekings, J. 2006. "Employment guarantee or minimum income? Workfare and welfare in developing countries". Basic Income Guarantee (BIG) conference, Philadelphia, 24–26 February. www.researchgate.net/publication/252273550_Employment_guarantee_or_minimum_income_Workfare_and_welfare_in_developing_countries/link/5852a dcb08ae0c0f32226e6c/download.

Selingo, J. 2018. "The false promises of worker retraining". *The Atlantic.* www.theatlantic.com/education/archive/2018/01/the-false-promises-of-worker-retraining/549398.

Shao, J. 1986. "The villagization program and the disruption of the ecological balance in Tanzania". *Canadian Journal of African Studies* 20(2): 219–39. doi:10.2307/484871.

Shapiro, J. 2001. *Mao's War Against Nature.* Cambridge: Cambridge University Press.

Sharma, N. 2006. *Home Economics.* Toronto, ON: University of Toronto Press.

Shellenberger, M. & T. Nordhaus 2004. *The Death of Environmentalism.* Oakland, CA: Breakthrough Institute.

Shotwell, A. 2016. *Against Purity.* Minneapolis, MN: University of Minnesota Press.

Sigley, G. 2006. "Chinese governmentalities: government, governance and the socialist market economy". *Economy and Society* 35(4): 487–508. doi:10.1080/03085140600960773.

Silver, N. 2012. *The Signal and the Noise.* New York: Penguin.

Simpson, L. 2017. *As We Have Always Done.* Minneapolis, MN: University of Minnesota Press.

Slorach, R. 2015. *A Very Capitalist Condition.* Berkeley, CA: Bookmarks.

Smith, L. 1999. *Decolonizing Methodologies.* London: Zed Books.

Smith, N. 1996. *The New Urban Frontier.* London: Routledge.

Sobel, A. 2014. *Storm Surge.* New York: Harper.

Soss, J., R. Fording & S. Schram 2011. *Disciplining the Poor*. Chicago, IL: University of Chicago Press.

Spence, M. 2000. *Dispossessing the Wilderness*. Oxford: Oxford University Press.

Spivak, G. 1999. *A Critique of Postcolonial Reason*. Cambridge, MA: Harvard University Press.

Srnicek, N. 2017. *Platform Capitalism*. Cambridge: Polity.

Sseviiri, H. *et al.* 2020. "Claiming value in a heterogeneous solid waste configuration in Kampala". *Urban Geography* 43(1): 59–80.

Standing, G. 2020. *Battling Eight Giants*. London: Bloomsbury.

Steinman, E. 2016. "Decolonization not inclusion: Indigenous resistance to American settler colonialism". *Sociology of Race and Ethnicity* 2(2): 219–36. doi:10.1177/2332649215615889.

Stern, A. 2016. *Raising the Floor*. New York: Public Affairs.

Stokes, K. & M. Lawhon (in press). "What counts as infrastructural labour? Community action as waste work in South Africa". *Area, Development and Policy*. doi:10.1080/23792949.2022.2145321.

Strong-Boag, V. 1979. "'Wages for housework': mothers' allowances and the beginnings of social security in Canada". *Journal of Canadian Studies* 14(1): 24–34. doi:10.3138/jcs.14.1.24.

Sutherland, C. *et al.* 2014. "Water and sanitation provision in eThekwini Municipality: a spatially differentiated approach". *Environment and Urbanization* 26(2): 469–88. doi:10.1177/0956247814544871.

Sundberg, J. 2014. "Decolonizing posthumanist geographies". *Cultural Geographies* 21(1): 33–47. doi:10.1177/1474474013486067.

Swyngedouw, E. 2015. *Liquid Power*. Cambridge, MA: MIT Press.

Symons, J. 2019. *Ecomodernism*. Cambridge: Polity.

Tcherneva, P. 2020. *The Case for a Job Guarantee*. Cambridge: Polity.

Thakholi, L. 2021. "Conservation labour geographies: subsuming regional labour into private conservation spaces in South Africa". *Geoforum* 123: 1–11. doi:10.1016/j.geoforum.2021.04.019.

Thompson, D. 2012. "The economic history of the last 2000 years: part II". *The Atlantic*. www.theatlantic.com/business/archive/2012/06/the-economic-history-of-the-last-2000-years-part-ii/258762/.

Thompson, M. 2022. "Money for everything? universal basic income in a crisis". *Economy and Society* 51(3): 353–74. doi:10.1080/03085147.2022.2035930.

Thoreau, H. 1854. *Walden*. Boston: Ticknor & Fields.

Tienhaara, K. & J. Robinson (eds) 2022. *Routledge Handbook on the Green New Deal*. Abingdon: Routledge.

Tindall, D., R. Trosper & P. Perreault (eds) 2013. *Aboriginal Peoples and Forest Lands in Canada*. Vancouver, BC: UBC Press.

Tokman, V. 2007. "Modernizing the informal sector". UN/DESA Working Paper No. 42. ST/ESA/2007/DWP/42. www.un.org/esa/desa/papers/2007/wp42_2007.pdf.

Tomiak, J. *et al.* 2019. "Introduction: settler city limits". In H. Dorries *et al.* (eds), *Settler City Limits*. Winnipeg, MB: University of Manitoba Press.

Torkelson, E. 2022. "Deserving and undeserving welfare states: cash transfers and hegemonic struggles in South Africa". *Journal of Southern African Studies* 48(1): 43–60. doi:10.1080/03057070.2022.2004772.

Tsing, A. 2015. *The Mushroom at the End of the World*. Princeton, NJ: Princeton University Press.

Tucker, J. & M. Anantharaman 2020. "Informal work and sustainable cities: from formalization to reparation". *One Earth* 3(3): 290–9. doi:10.1016/j.oneear.2020.08.012.

Vallas, S. & J. Schor 2020. "What do platforms do? Understanding the gig economy". *Annual Review of Sociology* 46(1): 273–94. doi:10.1146/annurev-soc-121919-054857.

Van den Bergh, J. 2011. "Environment versus growth – a criticism of 'degrowth' and a plea for 'a-growth'". *Ecological Economics* 70(5): 881–90. doi:10.1016/j.ecolecon.2010.09.035.

Van den Bergh, J. & H. Verbruggen 1999. "Spatial sustainability, trade and indicators: an evaluation of the 'ecological footprint'". *Ecological Economics* 29(1): 61–72. doi:10.1016/S0921-8009(99)00032-4.

Van Dijck, J., T. Poell & M. De Waal 2018. *The Platform Society*. Oxford: Oxford University Press.

van Parijs, P. & Y. Vanderborght 2017. *Basic Income*. Cambridge, MA: Harvard University Press.

Van Trier, W. 2021. "From 'Second Cheque Strategy' to 'Basic Income': Why Did André Gorz Change His Mind?" In P. Sloman, D. Zamora Vargas & P. Ramos Pinto (eds), *Universal Basic Income in Historical Perspective*. Cham: Palgrave Macmillan.

van Vliet, B., G. Spaargaren & P. Oosterveer 2010. *Social Perspectives on the Sanitation Challenge*. Berlin: Springer.

Vasudevan, R. & M. Novoa 2022. "Pluriversal planning scholarship: embracing multiplicity and situated knowledges in community-based approaches". *Planning Theory* 21(1): 77–100. doi:10.1177/14730952211000384.

Vergara-Camus, L. 2014. *Land and Freedom*. London: Zed Books.

Vosko, L. 2000. *Temporary Work*. Toronto, ON: University of Toronto Press.

Voss, J., D. Bauknecht & R. Kemp (eds) 2006. *Reflexive Governance for Sustainable Development*. Cheltenham: Elgar.

Wakefield, S. 2020. *Anthropocene Back Loop*. London: Open Humanities Press.

Walcott, R. 2009. "Reconstructing manhood; or, the drag of black masculinity". *Small Axe* 13(1): 75–89. doi:10.1215/07990537-2008-007.

Walenta, J. 2021. "The making of the corporate carbon footprint: the politics behind emission scoping". *Journal of Cultural Economy* 14(5): 533–48. doi:10.1080/17530350.2021.1935297.

Walia, H. 2021. *Border and Rule*. Chicago, IL: Haymarket.

Walker, G. & E. Shove 2007. "Ambivalence, sustainability and the governance of socio-technical transitions". *Journal of Environmental Policy & Planning* 9(3/4): 213–25. doi:10.1080/15239080701622840.

Walker, P. 2007. "Political ecology: where is the politics?". *Progress in Human Geography* 31(3): 363–9. doi:10.1177/0309132507077086.

Wallerstein, I. 1974. *The Modern World System*. San Diego, CA: Academic Press.

Webb, C. 2021. "Giving everyone a fish: Covid 19 and the new politics of distribution". *Anthropologica* 63(1): 1–16. doi:10.18357/anthropologica6312021275.

Weber, M. 1930. *The Protestant Ethic and the Spirit of Capitalism*. London: Allen & Unwin.

Weeks, K. 2011. *The Problem with Work*. Durham, NC: Duke University Press.

Weeks, K. 2020. "Anti/postwork feminist politics and a case for basic income". *tripleC* 18(2): 575–94. doi:10.31269/triplec.v18i2.1174.

Weheliye G. 2014. *Habeas Viscus*. Durham, NC: Duke University Press.

Whatmore, S. 2009. "Mapping knowledge controversies: science, democracy and the redistribution of expertise". *Progress in Human Geography* 33(5): 587–98. doi:10.1177/0309132509339841.

Whitaker, R. 2001. *Mad in America*. New York: Basic Books.

White, D. 2019. "Ecological democracy, just transitions and a political ecology of design". *Environmental Values* 28(1): 31–53. doi:10.3197/096327119X15445433913569.

White, D. 2020. "Just transitions/design for transitions: preliminary notes on a design politics for a green new deal". *Capitalism Nature Socialism* 31(2): 20–39. doi:10.1080/10455752.2019.1583762.

Whyte, K. 2017. "Indigenous climate change studies: Indigenizing futures, decolonizing the Anthropocene". *English Language Notes* 55(1): 153–62. doi:10.1215/00138282-55.1-2.153.

Wiedmann, T. & J. Barrett 2010. "A review of the ecological footprint indicator – perceptions and methods". *Sustainability* 2(6): 1645–93. doi:10.3390/su2061645.

Wimmer, A. 2002. *Nationalist Exclusion and Ethnic Conflict*. Cambridge: Cambridge University Press.

Winders, J. & B. Smith 2019. "Social reproduction and capitalist production: a genealogy of dominant imaginaries". *Progress in Human Geography* 43(5): 871–89. doi:10.1177/0309132518791730.

Wolford, W. 2010. *This Land Is Ours Now*. Durham, NC: Duke University Press.

Wood, E. 1999. *The Origin of Capitalism*. New York: Monthly Review Press.

Woodcock, J. & M. Graham 2020. *The Gig Economy*. Cambridge: Polity.

World Commission on Environment and Development 1987. *Our Common Future* (Brundtland Report). New York: United Nations.

Worster, D. 1994 [1977]. *Nature's Economy: The Roots of Ecology*. Garden City, NY: Doubleday.

Worster, D. 2016. *Shrinking the Earth*. Oxford: Oxford University Press.

Wright, E. O. 2004. "Basic income, stakeholder grants, and class analysis". *Politics & Society* 32(1): 79–87. doi:10.1177/0032329203261099.

Wright, E. O. 2012. "Transforming capitalism through real utopias". *American Sociological Review* 78(1): 1–25. doi:10.1177/0003122412468882.

Wright, E. O. 2019. *How to be an Anticapitalist in the Twenty-First Century*. London: Verso.

Wright, S. 2002. *Storming Heaven*. London: Pluto Press.

Wu, K. & J. Zhang 2019. "Living with waste: becoming 'free' as waste pickers in Chinese cities". *China Perspectives* 2019(2): 67–74. doi:10.4000/chinaperspectives.9184.

Wulf, A. 2015. *The Invention of Nature*. New York: Knopf.

Yang, A. 2018. *The War on Normal People*. New York: Hachette.

Young, I. 1986. "The ideal of community and the politics of difference". *Social Theory and Practice* 12(1): 1–26. doi:10.5840/soctheorpract198612113.

Zanotti, L. *et al.* 2020. "Sustainability, resilience, adaptation, and transformation: tensions and plural approaches". *Ecology and Society* 25(3): 4. doi:10.5751/ES-11642-250304.

Zapata Campos, M. & P. Zapata 2013. "Switching Managua on! Connecting informal settlements to the formal city through household waste collection". *Environment and Urbanization* 25(1): 225–42. doi:0.1177/23996544221118191.

Zhao, J. & L. Whitehead 2022. "A guaranteed basic income could end poverty, so why isn't it happening?". *The Conversation*. https://theconversation.com/a-guaranteed-basic-income-could-end-poverty-so-why-isnt-it-happening-182638.

Zuboff, S. 2019. *The Age of Surveillance Capitalism*. New York: Public Affairs.

INDEX